ORGANIC FUTURES

Organic Futures

The Case for Organic Farming

Adrian Myers

Green Books

First published in the UK in 2005
by Green Books Ltd,
Foxhole, Dartington,
Totnes, Devon TQ9 6EB

Copyright © 2005 Adrian Myers

All rights reserved

No part of this book may be used or reproduced in any manner without written permission, except in the case of brief quotations in critical articles or reviews.

Cover design by Rick Lawrence
samskara@onetel.net.uk

ISBN 1 903998 69 7

Printed on Five Seasons Stone-White 100 per cent recycled paper
by MPG Books, Bodmin, Cornwall, UK

Contents

Chapter 1	What is Organic Farming?	9
Chapter 2	From the Past to the Present	34
Chapter 3	The Living Soil	65
Chapter 4	Nutrition and Food Safety	87
Chapter 5	Genetically Modified Foods	107
Chapter 6	Maintaining Soil Fertility	132
Chapter 7	Control versus Co-operation	155
Chapter 8	The Challenges Ahead	181
Chapter 9	The Organic Uplands: A Vision of the Future	210
Notes		239
Bibliography		244
Useful Organisations and Websites		245
Index		246

DEDICATION

To my friend John,
who encouraged me to write this book
in the first place

To my wife of thirty-five years,
who never once asked me to
go out and get a "proper job"

&

To the Divine in all of us

Chapter 1

What is Organic Farming?

"The organic movement is clearly established and here to stay. It is heralding a change in agriculture which is occurring simultaneously in every developed agricultural nation in the world. Far from being a return to the past, organic farming is an agriculture of the future, our future."—Nicolas Lampkin, *Organic Farming*[1]

Ancient and Modern

On a cold, grey, drizzly day I slipped into the warmth of the Swiss Cottage reference library in London to continue my researches into organic farming and horticulture. It was 1971, and the day I discovered two amazing things: the fact that the Chinese, Japanese, and Koreans had been practising organic farming for over forty centuries non-stop, without a loss of soil fertility;[2] and a quote from the original statement of the Soil Association (the leading organic organization in Britain), which read as follows:

"Many scientists and agriculturists now realize that their knowledge of the natural processes underlying soil fertility is incomplete. They recognize that these processes are only partly explicable in terms of agricultural chemistry and that the pure inorganic approach to the study of soil science is a line of thought as dead as the mechanical determination of nineteenth-century physics. 'Dead' is the appropriate word, for the missing link is life itself."

The discovery that some countries' farmers had been farming organically for 4,000 years was revelation enough, and showed conclusively that organic husbandry had a long and noble history; but to discover that modern organic farming is very much more than just farming as practised in this country before 1939, as so many think, really made me sit up and take notice. I began to realize that organic farming is a changing and evolving science, constantly being moved forward by the growing knowledge of the biological and ecological sciences, and as such is at the cutting edge of agricultural development. Conventional farming, on the other hand, began to seem increasingly old-fashioned and limited in its approach. I was becoming aware that not only does organic farming have an enduring past, but it also has a certain future, as the quote at the beginning of this chapter indicates. I would add one caveat to Nicolas Lampkin's statement: that organic and sustainable farming is also growing fast in many Third World countries.

In 1971 my new wife and I were living in London, but had made a major decision to give up our business and find a small farm in the country which we were going to farm organically. In between rushing off to Devon, Herefordshire, Wales and Shropshire on property hunts, we were busy learning everything we could about the subject. In the process we discovered the great classical books of the organic movement. We spent our time avidly taking notes as we read Sir Albert Howard's *An Agricultural Testament*, Eve Balfour's *The Living Soil*, Newman Turner's *Fertility Farming*, Lawrence D. Hills' *Grow Your Own Fruit and Vegetables*, Masanobu Fukuoka's *One Straw Revolution*, Edward Faulkner's *Ploughman's Folly*, and of course F. H. King's *Farmers of Forty Centuries: Permanent Agriculture in China, Korea and Japan* (now republished with the subtitle *Organic Farming in China, Korea and Japan*). This last provided us with all the evidence we needed that organic husbandry really does work. There are many experiments with organic farming and horticulture being done around the world, but it should be enough for anyone that the longest trial in history has proved itself. It is possible for modern Chinese to read ancient records going back to the earliest times, and because there has been an almost unbroken civilization over this period, with agricultural records going back that far in some provinces of China, we know that soil fertility and crop yields have remained fairly consistent over this period.[3]

More enlightened members of past societies, like China, through intuitive cognition, observation and experiment, developed the concept of the 'Law of Return'. They noticed that all living things, both plant and animal, would at some stage die and rot, and that the resultant product encouraged the growth of the next generation. This is the continuous cycle of birth, growth, death and rebirth. In the autumn, deciduous leaves fall to the ground, rot down and provide nutrition for future growth. Grass where animals have defecated or urinated grows greener and is more luscious. They discovered that if they combined all the waste organic material together away from the field in compost heaps, it produced more healthy and productive crops, it improved the soil structure, and over time could even increase both the long-term fertility and the depth of the living topsoil. As the ancient saying goes: "Corruption is the mother of vegetation."

It was noticed that some tougher crop residues rotted down more readily with the addition of animal manure and urine, and so they developed a sophisticated system of returning all organic waste back onto the land by first piling it in huge heaps which rotted down over the season, ending up as an easily manageable and sweet-smelling product that was pleasant to handle. They rescued everything they could to add to their heaps. They realized that goodness in the soil ran off into their drainage ditches, so they collected the silt and harvested water hyacinth and other plants growing in the ditches and recycled them as well, adding them to the compost heaps. Everything that was organic and would rot was recycled back onto the land, with the result that a vibrant, living and healthy soil ecosystem was maintained that was capable of producing sustainable yields and healthy crops for as long as these practices were continued.

As Sir Albert Howard, one of the pioneers of the modern organic movement, observed: "The maintenance of the fertility of the soil is the first condition of any permanent system of agriculture."[4] It is only in comparatively modern times that soil scientists have begun to understand the living soil and the natural processes involved in plant nutrition, but ancient farmers intuitively understood this. As F. H. King noted:

> "While it was not until 1877 to 1879 that men of science came to know that the processes of nitrification, so indispensable to agriculture, are due to germ life, in simple justice to those who through all ages from Adam down, living close to Nature and working through her and with her, have fed the world, it should be recognized that there have been those among them who have grasped such essential, vital truths and have kept them alive in the practices of their day."[5]

The Aim of this Book

Looking back at history is a very valuable exercise, for it shows us what is possible. However, the main purpose of this book is to show that only sustainable agriculture and horticulture have any sort of future when it comes to food production. By definition, only sustainable forms of farming and horticulture will continue to survive, by whatever name you like to call them, whether 'biological', 'ecological', 'sustainable', 'fertility', 'biodynamic' or 'organic'. On the other hand, a system of husbandry that has a tendency to denude and pollute the natural fertility of the soil, is by definition without a long-term future. I am fully aware that there is little chance of all agriculture becoming strictly organic, but it has to become sustainable. I use the term 'sustainable' in a very specific way, to mean sustainable over time, maintaining or even improving true biological soil fertility; and sustainable in the sense of not degrading and poisoning the environment in the process. Not all sustainable systems are strictly organic, but I have concentrated on organic husbandry as the standard to aim for. Whilst putting the case unequivocally for sustainable and organic husbandry, however, I have tried to avoid the trap that many proponents of organic farming often fall into, namely self-righteousness. If I do lapse occasionally, my excuse is enthusiasm. As Jules Pretty, an advocate of sustainable farming, has quite rightly cautioned, those who are concerned with the development of a more sustainable agriculture must not fall into the same traps by making grandiose claims to have the sole answer.

This is not a handbook on how to farm and garden organically; there are many excellent books on the subject, such as Nicolas Lampkin's *Organic Farming* and Bob Flowerdew's *Organic Bible: Successful Gardening the Natural Way*. Nor is this a consumer guide, because this subject is covered comprehensively in Lynda Brown's indispensable *The Shopper's Guide to Organic Food*.

What it does is attempt to show organic husbandry as continuously evolving and developing, through the innovative developments of organic farmers and advances brought about by scientific research and understanding. Organic husbandry is not about returning to farming and gardening as it was seventy to a hundred years ago; indeed, I aim to show that the organic approach is more modern in its outlook and practice than conventional husbandry, which is stuck in the 'modernistic' paradigm based largely on an oversimplification of Liebig's discoveries—the view that plants require nutrition solely in the form of water-soluble chemicals. The modern organic approach, on the other hand, is based on an evolving knowledge of soil ecosystems as understood by the biological sciences. I argue that intensive, sustainable approaches to growing food are essential if we are to feed the world whilst protecting the environment, creating a truly fertile and thriving soil ecosystem, and at the same time protecting our own health, not only for our own generation but for those to come. I contend that so-called 'conventional' agriculture, which uses huge amounts of energy to fuel it, inevitably has a limited future, and that any form of agricultural system that considers the pollution of our food and environment as a price worth paying for progress, is fundamentally flawed.

Running as an undercurrent throughout this book is the story of the living soil itself. Chapter Three seeks to explain the fascinating world of the soil ecosystem and the interrelated and co-operative interactions of plants, micro-organisms, small animals, and fungi that make up the living rhizosphere (the topsoil and its environment), which has maintained life on this earth for millions of years, and which, when nurtured and enlivened, can provide us with all the healthy food we will ever need.

The definition of any sustainable system of agriculture, and indeed a sustainable society, is the extent to which it recycles its waste, in particular its organic waste. Chapter Six discusses recycling systems, with examples from around the world, the theory and the composting of bio-waste—from the smallest heap to municipal and company composting systems on a large scale.

Inevitably I also discuss the problems caused by conventional farming's use of fertilizers, pesticides and intensive animal production, as well as BSE and genetically modified (GM) crops. Above all, how did we come to be farming in this way in the first place?

I try to challenge the reader by showing that, as with many aspects of modern living, we are faced with a choice not between going backwards or forwards, but between two alternative futures: one based on life-harming activities and the wish to dominate Nature, the other based on life-sustaining activities and the recognition that we are part of Nature. It is our changing attitude to Nature and our relationship to it that is another undercurrent running throughout this book. This attitude is seen as the reason for the way we practise agriculture today, and I explore how by re-evaluating our relationship to the rest of Nature—and our own nature—we can re-evaluate our attitude to the land, and therefore the way we grow our food.

To finish on a positive vision of the future, the last two chapters discuss how agriculture can be regenerated in both the West and the Third World, with many inspiring examples of successful and sustainable programmes that illustrate the huge growth of organic and sustainable methods throughout the world.

I should like to explain my use of the terms 'agriculture' and 'farming'. When discussing organic agriculture, the subject is often restricted to the Western, commercial way of farming. This, in my view, is a very exclusive and limited way of thinking. I use the term to include all farming, whether small- or large-scale, horticulture, forestry, or private or community gardening for food. In the Third World, as Vandana Shiva (physicist, ecologist and activist) has so forcefully commented:

> "It is women and small farmers working with biodiversity who are the primary food providers in the Third World, and contrary to the dominant assumption, their biodiversity-based small farms are more productive than industrial monocultures."[6]

Above all, I hope the book will be of interest to the general public, who are becoming increasingly interested in the subject not only because of concerns about the quality of our food, health, the environment, GM crops and the treatment of animals, but also because they want to know more about the philosophies, science, ideas and history of organic methods over the centuries, as well as the latest scientific knowledge that has transformed our understanding of the ancient methods of farming and taken them to new levels.

Defining Organic Farming

When we go to the shops to buy something marked 'organic produce', what does it mean? What are organic farmers doing that conventional farmers are not? We often think that organic farming is just about growing crops without chemicals, fertilizers, pesticides, herbicides and fungicides, but in fact it is much more profound than that. The best way to start is to begin by quoting from Lynda Brown's *The Shopper's Guide to Organic Food*:

> "In organic farming systems, food production is viewed not as the supplying of a commodity, but as a holistic enterprise where sustainability and health, each important as the other, are interlinked, and where healthy food can be produced only on healthy soils and from animals reared on a natural diet."[7]

This is all well and good, but let us broaden the picture. There is a profound difference between the philosophical and even scientific views of organic and conventional farmers, mainly in their attitudes towards Nature in the broadest sense, but also in their views on the natural ecology of the soil and on plant nutrition.

Mother Earth—Gaia

Traditional societies had, and still have, a concept of 'Mother Earth'—a living entity, self-organizing, self-sustaining, self-regenerating and self-regulating. This leads to a reverence and respect for Nature, of which those societies and individuals saw themselves an integral part. What happened to European thought over the centuries, and particularly during the eighteenth and nineteenth centuries, was a change in this perception. As Vandana Shiva has noted: "The transformation of the perception of nature during the industrial and scientific revolutions illustrates how 'nature' was transformed in the European mind from a self-organizing, living system to a mere raw material for human exploitation, needing management and control."[8]

This ancient idea of Nature as intrinsically interconnected and self-organizing was re-awakened with the development of the science of ecology, and the study and increased understanding of the biosphere (the totality of all the organisms on Earth) and our planetary system by people such as the scientist James Lovelock in the 1970s. He took the interconnectedness of species and the environment a stage further, seeing the world's environment and the total life of the planet as functioning in a coherent way, as functioning as a self-regulating single super-organism, which he called Gaia.

Unfortunately, Lovelock's theories were misunderstood by both the scientific community and some of his supporters. The fact that he was championed by many 'new age' thinkers only enhanced the suspicions of the scientific community, although he has always been very clear to qualify his theory:

> "I have frequently used the word Gaia as a shorthand for the hypothesis itself, namely that the biosphere is a self-regulating entity with the capacity to keep our planet healthy by controlling the chemical and physical environment. Occasionally it has been difficult, without excessive circumlocution, to avoid talking of Gaia as if she were known to be sentient. This is meant no more seriously than is the appellation 'she' when given to a ship by those who sail her, as the recognition that even pieces of wood and metal when specifically designed and assembled may achieve a composite identity with its own characteristic signature, as distinct from being the mere sum of its parts."[9]

Despite its initial scepticism, the scientific community has recently begun to take Lovelock's theories more seriously. His theory of a 'self-sustaining super-organism' has been re-evaluated in the light of modern complexity theory.[10] This mathematical theory deals with the way a spontaneous transition occurs from a less orderly to a more orderly state in the internal dynamics of complex systems when they reach a critical level of complexity. For example, a computer simulation of the internal interactions in an ant colony, in which the ants obey only a few simple rules, revealed rhythmic and orderly patterns of behaviour within the colony—and these patterns emerged purely as a property of the system itself. Interestingly, complexity theory also applies to non-living systems. This means that Lovelock's idea that both the Earth's biological life forms and its non-life components, such as the atmosphere, interact in a coherent and self-sustaining way, begins to make mathematical and scientific sense.

There are those scientists who see Darwinian natural selection as working exclusively on the individual level and not on the group level. However, those who understand complexity theory would argue that there is no contradiction between the two approaches. The individual selfishly works for its own ends, but unwittingly functions in a co-operative and coherent way with others in the group, to the benefit of both the individual and the group—and this all occurs in a non-purposeful way, merely as a function of the internal dynamics of the group.

For an example of Nature as self-regulating, we only have to look at the Earth's atmosphere.[11] As LOvelock has noted, the mixture of gases in the atmosphere cannot possibly be maintained by purely physical and chemical processes. Gases such as methane, nitrous oxide and even nitrogen should not survive in the presence of oxygen, but they do—and the presence of oxygen itself is an anathema. Reactions should have occurred over time which would have led to a state of equilibrium, but the constituents of the Earth's atmosphere have been kept in a state of disequilibrium for millions and millions of years. Our earth's atmosphere has been created, regulated and maintained at the optimum mixture by the myriad forms of life that make up the biosphere of this amazing planet. Oxygen itself was created by early life forms, and the present balance of oxygen and ammonia in the atmosphere is maintained in its present optimum ratios by life itself. As Lovelock has noted, "The climate and the chemical properties of the Earth now and throughout history seem always to have been optimum for life. For this to have happened by chance is as unlikely as to survive unscathed a drive blindfold through rush-hour traffic."[12]

What life on Earth does, through a process of feedback mechanisms, is to maintain relatively constant conditions on the planet by a series of active controls or homeostasis. But it is more than just life: it is the complex interaction of the physical environment, the atmosphere and the oceans, as well as life itself, that can be seen as Nature's maintenance of the Earth's environment. This vision of Nature, of which we are part, is awe-inspiring.

The view of Nature as self-organizing, self-sustaining, self-regenerating and self-regulating is at the heart of the organic and sustainable approach to agriculture. It means that the farmer has faith in natural processes to nurture plants and animals in the same way that it has been doing for millions of years. Of course the organic farmer sows the crops and keeps the animals he or she finds most useful for food production, but it is done in a co-operative and respectful frame of mind. It is a recognition that no gardener has ever 'grown' a plant, no farmer has ever 'produced' an animal. The best we can do is create the right conditions for them to grow themselves.

A healthy soil is teeming with life—bacteria, actinomycetes, fungi, algae, microscopic plants, protozoa, tiny enchytraeid worms, earthworms, arthropods, and a whole range of animals. Through complicated processes, described in greater detail in Chapter 3, all this life makes food available for the plants, and ultimately for animals and us. So farmers, and human beings in general, cannot be other than part of Nature.

Organic Farming in Practice

How, then, does this vision express itself in the practice of organic husbandry?

Integrated Biodiversity

A truly organic farm has to be as mixed and diverse as possible, with a mixture of animals and crops, along with woodland and other varieties of habitat, ideally including wildlife areas.[14] And, most importantly, all the different practices should be part of an integrated whole. No activity, crop or animal should be seen as a separate enterprise; everything on the farm is interrelated and supports all the other enterprises.

This principle was well understood in the past, largely because of the necessity of maintaining soil fertility. For instance, in the eastern counties of England, the emphasis was on growing grain, as it still is. But today, large areas of south-east Yorkshire, Lincolnshire, Cambridgeshire, Norfolk and Suffolk grow grain crops on a regular basis, year after year, with occasional breaks growing other arable crops. Before the advent of chemical fertilizers and pesticides, even these 'breadbaskets' of England had forms of mixed farming, using sheep and working horses to help maintain the soil's fertility through their manure. Grain crops were alternated with clover and root crops, on which the sheep grazed. This 'Norfolk four-course rotation system', as it was called, and the many varieties of it, kept the fertility of those soils at a level where grain could continue to be grown without a loss of fertility.

In the 1980s I gave a talk to a local organic gardening group in Staffordshire about how we ran our farm. The theme of the talk was about the interrelatedness of all the different parts of the farm, and how this mirrored the way in which all parts of the natural world interrelate—or as the German poet and philosopher Goethe said, "Nothing happens in living nature that is not in relation to the whole." For example, on our farm we grew a commercial crop of potatoes and kept pigs. Inevitably, there were potatoes that were injured, or otherwise unsaleable. These were cooked and used as a carbohydrate substitute in the pig's feed, saving money. The pig manure and bedding straw were important ingredients in the compost which helped to grow the potatoes. The bedding straw was straw from the crops of wheat, which also helped to feed the pigs. It was also used as bedding for the cows, whose by-products from butter production also helped to supplement the pigs' rations, and whose manure and bedding straw helped to produce good compost, which helped to grow a perennial crop of Russian comfrey, which also helped to feed the pigs and provide a valuable high-potash liquid manure, which was used to grow tomatoes, courgettes, squash and soft fruit which we sold once a week in Shrewsbury market—and so on—you get the picture.

Of course, there will be this sort of integration at some level on most farms, but on an organic system there tends to be a much greater level of diverse enterprises, and their interdependence is far more important for both economic reasons as well as the maintenance of soil fertility.

There is another vitally important reason for creating as much biodiversity on a farm as possible—there is a direct correlation between the biodiversity of a system and its overall health. Ecologists were the first to recognize how important it is that any natural environment has as wide a range of plants and animals as possible. In fact they went further, in realizing that they could measure the sustainability and stability of an ecosystem purely by the number of species recorded in a given area. When the species are reduced to only a few types, the system becomes fragile and unhealthy. On the other hand, an environment that is species-rich, with a high genetic diversity, is very resilient. This strength and resilience comes from its diversity and flexibility. The same is true of the environment of a farm. Where there is a wide range of crops, animals, different varieties of grasses, clovers and other useful 'herbs' in a pasture, and a wide range of habitats and a healthy soil with a complicated mix of micro-organisms in it, the farm as a whole is a healthier and more productive place, with fewer pests and diseases.[15]

Feed the Soil, not the Plant

This leads very neatly to the very basis of all this activity, indeed the engine of the farm itself—the living soil. All the soil micro-organisms, soil flora and fauna, and indeed all the plants and animals that rely on it, have to be encouraged and enhanced, because the living components of the soil are essential for the nutrition and good health of the plants that grow in it, and the animals and humans that feed on them.

Ever since plants started to grow on the land millions of years ago, they have had two sources of food: the air, from which they get carbon dioxide and water, which they convert to carbohydrates and sugars providing their energy and cellulose, which in turn provides the structure of the plant itself; and the soil, which provides nutrients in the form of microscopic bacteria, algae and fungi.[16] On a conventional farm, the plants' nutrients are largely supplied in the form of water-soluble chemical fertilizers produced in factories. On an organic farm, however, the farmer's soil is fed with rotted organic waste products, which the soil organisms convert into soluble nutrients for the plants. On a conventional farm the natural system is bypassed, which leads to many negative consequences. On an organic farm the natural system is encouraged and nurtured, which leads to healthier crops, animals and environment, and the creation and maintenance of a healthy soil.

The 'Law of Return'

With organic waste materials, the principle is 'recycle everything'. The aim is always to be as self-sufficient as is possible when it comes to feeding the soil. It is perfectly possible for the farm to supply all the needs of both the soil and the crops from within the farm itself, with the exception of soil conditioners like lime. However, where it is necessary, the Soil Association's rules allow organic farms to import organic waste onto the farm, as well as certain types of waste from non-organic sources as long as they are composted first; but no manure from intensive animal farms is allowed. Another way of importing nutrients onto the farm to speed up the revitalization of the soil, especially in the early years after conversion to organic practices, is in the form of animal food which ends up as manure. Also, in intensive horticultural systems, where it is not always possible to provide adequate crop nutrition from within the system, imported organic vegetable and animal waste can be used to maintain soil fertility. The Soil Association permits the use of imported organic waste and a restricted amount of organic waste from non-organic sources.

Apart from the exceptions above, the aim of the sustainable and organic farming movement is to be as self-sufficient as possible in its supply of soil nutrients. This is made possible by:

1. Recycling all organic matter on the farm.

2. Relying on the ability of plants to utilize carbon dioxide from the air with the help of sunlight, through the process of photosynthesis, to provide their energy and physical structure—thus continuously replenishing organic matter so essential for the creation and maintenance of soil fertility.

3. Growing crops like clover, which have bacteria called Rhizobia which live in adapted nodules on the roots of the plants and have the ability to fix atmospheric nitrogen, which helps to maintain soil nutritional levels.

With these processes, soil fertility can be improved and maintained indefinitely. After all, this is all that Nature has been doing for millions of years, without our input. If the natural system had not been self-sufficient and sustainable, it would have ceased to exist long ago. Only farming systems that are sustainable have any credibility, and indeed any long-term future over time. Moreover, the system should be environmentally sustainable. Destroying soil fertility and ecology, spending huge amounts of fossil fuels to make chemicals and to transport them over huge distances, and then having to spend large amounts of money cleaning up the pollution caused by poisons and fertilizers, is not by any stretch of the imagination sustainable.

Respect for Animals

According to surveys, when it comes to the production of food, one of the most important concerns of the general public is animal welfare. The international standards for organic farming make the principles very clear: "To give all livestock the conditions of life that allow them to perform all aspects of their innate behaviour." This means that organic farmers should aim to keep their animals as naturally as possible within the constraints of a farm.

For example, chickens were originally wild jungle fowl from India, but this does not mean we have to recreate jungle conditions for them. What it does mean is that they should go outside, that they have grass and other green food to eat, that they can scratch the ground to hunt for worms and other insects, and that they have housing to shelter in at night or in poor weather. They also need somewhere to perch at night off the ground, which they will do instinctively given the chance, as they would do in the wild, in the branches of trees. Animals are often housed indoors on organic farms, but not permanently, which is prohibited under Soil Association standards. They should have fresh air, plenty of space to move around, and plenty of comfortable straw bedding which is changed regularly—in other words the highest standards of animal husbandry. Indeed, another of the Soil Association standards states: "Winter housing must be available for all livestock where severe weather or unhealthy or slippery ground conditions occur."

The International Federation of Organic Agricultural Movements (IFOAM)

To sum up this section on the practical application of the organic philosophy, one can do no better than quote from the International Federation of Organic Agricultural Movements (IFOAM) list of standards:

1. To produce food of high nutritional quality in sufficient quantity;
2. To work with natural systems rather than seeking to dominate them;
3. To encourage and enhance biological cycles within the farming system, involving micro-organisms, soil flora and fauna, plants and animals;
4. To maintain and increase the long-term fertility of soils;
5. To use as far as possible renewable resources in locally organized agricultural systems;
6. To work as far as possible within a closed system with regard to organic matter and nutrient elements;
7. To give all livestock the conditions of life that allow them to perform all aspects of their innate behaviour;
8. To avoid all forms of pollution that may result from agricultural techniques;

9. To maintain the genetic diversity of the agricultural system and its surroundings, including the protection of plant and wildlife habitats;

10. To allow agricultural producers an adequate return and satisfaction from their work including a safe working environment;

11. To consider the wider social and ecological impact of the farming system.

Conventional Farming

As we did with organic farming, let us first take a broad look at the principles and assumptions that underlie the practice of conventional farming. The differences between the two approaches should not be seen in terms of science versus non-science. Both organic and conventional systems of agriculture use science to further their knowledge and to improve their approaches. Both systems would argue that science has been instrumental in developing their versions of farming as it is practised today. However, as the two approaches start from very different underlying premises, the form that the scientific research takes, and the conclusions that are drawn from that research, will inevitably be very different.

From an Exponent of Conventional Agriculture

Here I attempt to argue from the modern conventional farmer's viewpoint, in order to give an insight into the conventional view of modern agriculture, and the direction its proponents see it heading.

"Agricultural practice has been transformed and improved beyond recognition over the last hundred years, with yields improving dramatically around the world, and this agricultural progress has come largely as a result of science, especially chemistry.

"The great discovery of plant nutrition and chemical salts was achieved by chemists like Liebig in the 1840s. He discovered the three most important plant foods—nitrogen, potash and phosphate—and published his famous work on plant nutrition. Or as the article on the history of chemical fertilizers in the *Encyclopaedia Britannica* has it: 'The milestones in the development of understanding of plant growth and nutrition rank in significance to mankind with the greatest achievements in chemistry, physics, mathematics, astronomy and other disciplines.'[17] Now, with a combination of soil testing and computer programmes one can find out exactly how much fertilizer should be applied, in what combination, at what time, and for what crop.

"Along with these innovations there have been—especially since 1945—valuable developments in controlling the pests and diseases that have plagued mankind throughout history, through the use of effective fungicides and insecticides. As a result, the combination of increased yields and effective disease and pest control has revolutionized agricultural production worldwide.

"In the UK there are over 350 pesticides to choose from. As a result, farming is much cleaner, neater and more precise than it used to be, with the guesswork taken out of it. It is now a scientific business, in which the farmer has almost complete control.

"Farming is a constant warfare against weeds, diseases and pests, and this is where the chemist comes in to help the farmer keep on top of nature. There are specialist herbicides which kill off weeds and leave a clean crop with no competition. There are insecticides and fungicides with spray programmes especially designed for each crop, so that the farmer knows exactly at which stage in the growth of the crop to apply which spray treatment. For every crop there is a scientific programme to follow for maximum results.

"The production of meat and dairy produce has also been made highly scientific and economic, with the maximization of egg production, liveweight gain in meat production, and milk production per cow. Through scientific breeding programmes and a greatly improved understanding of nutrition, production has increased dramatically over the years. Drugs, antibiotics and growth-promoting agents can be added to animal feed to help prevent them from becoming ill, and to speed up their growth rates to produce cheap food.

"Modern farming is big business, and is not called the 'farming industry' for nothing. Obviously the whole purpose is to maximize crop and meat production for maximum profit in the most efficient way. Competition dictates that farmers have to squeeze every last drop out of the system because the profit margins are getting smaller and smaller. To achieve this they use the best techniques and knowledge that science and business can provide.

"The result of this scientific revolution in farming practices over the last sixty years has been to increase yields and to produce crops and animal carcasses of consistent quality, which are at the same time as cheap as possible. As a result of these developments, food production per capita worldwide has risen by 7 per cent and in Asia by 40 per cent since the 1960s. New developments such as genetically modified foods have the potential of increasing yields still further as well as eliminating pests and diseases even better than in the past.

"In 1993 this scientific revolution in agriculture prompted Donald Plunkett, scientific adviser to the Consultative Group on International Agricultural Research, to describe this transformation as 'the greatest agricultural transformation in the history of mankind, and most of it has taken place in our lifetime. The change was brought about by the rise of science-based agriculture which permitted higher and more stable food production, ensuring food stability and security for a constantly growing world population.'[18]

"Over this period, agricultural scientists have spread this knowledge throughout the world. This has been part of a compassionate process of helping the less developed nations of the world, by showing how through using the great discoveries of the West, everyone can improve their agricultural production, eliminate pests and diseases—or at least keep them within manageable proportion—feed the hungry, and bring everyone's agricultural production techniques into the twenty-first century.

"Scientists increasingly feel misunderstood, often finding it difficult to explain their work to a public who fear change and can never fully understand scientific methods and practice. Increasingly among scientists there is a feeling that the scientific community and science itself is under siege.

"To move in the direction of organic farming would be a reversal. There is no way organic farming can even begin to feed the world. As Norman Borlaug, the director of Mexico's OEE (Office of Special Studies) said in 1992:

> 'Some agricultural professionals contend that the small-scale subsistence producers can be lifted out of poverty without the use of purchased inputs, such as modern crop varieties, fertilizer and agricultural chemicals. They recommend instead, the adoption of so-called "sustainable" technologies that do not require fertilizers and improved varieties. . . . The advent of cheap and plentiful fertilizers has been one of the great agricultural breakthroughs of humankind. . . . The adoption of science-based agricultural technologies is crucial to slowing—and even reversing—Africa's environmental meltdown.'[19]

"The kind of criticisms that scientists are currently experiencing are paralleled by the experience of farmers. After the Second World War, they were encouraged to modernize and produce cheap food as efficiently as possible.

Now they are being criticized for doing just that. Commodity prices have collapsed over the last few years, which has led to a serious economic crisis in farming, but 'cheap food' is still the call. So fields have been made bigger, hedges and trees have been grubbed out, ponds filled in and wet areas drained. All these improvements have, until recently, been subsidized by special grants. To be able to feed the population, industrialized agriculture is needed and modern agricultural science will continue to provide new progressive developments."

Brave New World
This, then, is the vision of pragmatic farmers and agricultural researchers not only in the West, but also in many countries in the Third World. This vision of science-based agriculture is but a part of a more inclusive modern scientific paradigm, itself in large part the result of a gradual change in the Western view of Nature and our relationship to it.

Our attitudes towards Nature in the West have changed fundamentally over the centuries, from seeing ourselves as an integral part of Nature, to seeing Nature as wild, untamed and open to improvement. This change, which had been developing throughout the Renaissance in Europe, shifted dramatically during the eighteenth-century 'Age of Enlightenment'. Thinkers at this time increasingly perceived the reasoning powers and creative activities of mankind as distinct from the 'natural world', which in turn was increasingly being seen as the product of random events based on intrinsic natural laws. Human civilization, 'civil' behaviour and the power of the human intellect and ability to reason were seen from this time onwards as superior to, and worlds apart from Nature.[20]

As a result, there was a growing belief that the human quest and destiny was to 'improve' on Nature and make a better world through the application of science and technology. As we have already seen, it was Liebig's discovery of the three main chemical foods that are essential for plant growth, that exemplified this trend. However, Liebig himself saw his discoveries in a much broader ecological and biological context. Unfortunately, the consequent application by agriculturists of his discoveries was based on a gross oversimplification of the understanding of plant nutrition, which still underlies the 'modern' approach to agriculture.

As Vandana Shiva has commented, Nature and the soil began to be seen as 'passive-female' through the growth of a patriarchal perception of Nature. Man knew best—the soil needed improving with applications of plant nutrition in the form of a very simple mix of water-soluble nitrates, phosphates and potash. The remarkable increases in the growth of crops that followed the early applications of these chemical fertilizers only confirmed the efficacy of this approach by both its supporters and many of its sceptics. What was not understood at the time was that these results were only achieved through the combination of existing natural fertility and the new chemistry. Later, when the natural fertility of the soil diminished through over-reliance on chemical fertilizers, their application had to be increased in order to maintain yields. The concept of 'passive' earth was being confirmed, and the continued use of chemical fertilizers became accepted as necessary.

By this time the bandwagon was well and truly rolling. Here was a vision of a truly 'brave new world', in which abundant crops could be grown indefinitely to feed the growing populations around the globe, through the simple application of chemistry. The possibility that crop and animal diseases could at last be eliminated with the new generation of powerful pesticides and antibiotics that were becoming available only increased the sense of optimism about the powers of humanity to outwit Nature and get the better of her.

Even in mixed-farming areas, such as in the part of Shropshire in which I live, animal manures are seen largely in terms of the direct nutrition they supply to a growing crop. In fresh animal manure there is a good proportion of available plant nutrients, but these are seen only as supplementary to the main chemical supply. This is a very different view from seeing organic waste as the valuable food for soil micro-organisms, which convert it into plant nutrients as and when they are needed, for the growing crop. In more intensive forms of modern farming, organic waste such as straw from crops or manure from intensively farmed livestock is seen as an unfortunate 'waste' that has to be disposed of. In these oversimplified systems, there is no way the 'waste' can be used. Even in a situation where these waste products can be recycled back onto the land, the effectiveness of soil micro-organisms has often been reduced so much that their ability to convert the organic matter into useful plant nutrients is restricted.

An exception to this rule is on non-organic farms where zero-tillage, which is becoming more and more popular, is practised. On these farms, there is no ploughing or tillage at all and the straw from each crop is mechanically chopped and left to rot on the ground. The seed drills have cutters on the front, which cut a slice through the crop residue and the top few inches of the soil to enable the seed to be planted. There are some very large grain farms in Argentina that have been using this method for twenty years or more, and the topsoil is increasingly rich in organic matter and humus, and it is reported that despite the chemical fertilizers and sprays used, the soil life has improved, and the crop yields have increased over time.[21] However, on organic farms in Latin America, which are using the same methods, the results are even more spectacular.

The Cost

There is no doubt that the application of chemical fertilizers and pesticides has led to an increase in agricultural production worldwide, but at what cost? The costs are on many levels: soil degradation, pollution of water supplies, the environment and our food, abuse of animals in intensive conditions, unsustainable uses of energy, and the increasing power of the large seed and chemical companies.

Soil Degradation

Soil degradation is one of the most damaging costs of modern farming methods in many ways: it means that whilst this generation literally rips off the land in order to feed itself, we are destroying the soil's fertility for future generations. The damage stretches from minor forms, which can be reversed over a few years or more of remedial organic farming methods, to permanent losses of topsoil due largely to a loss of shelter belts, and more importantly to the reduction of soil organic matter which makes soils more vulnerable to both water and wind erosion. It is estimated that in the past fifty years, the world has lost a quarter of its topsoil; the United States alone has lost a third. It has been estimated by Bob Evans that some 6,200 sq km (2,400 sq miles, 4.4 per cent) of soil in Britain is under high risk of erosion, and some 2,100 sq km (810 sq miles, 1.5 per cent) is under severe risk, due largely to 'modern' farming methods.[22] All around the world, soil erosion is an increasingly serious problem. To quote Franklin D. Roosevelt: "The nation that destroys its topsoil destroys itself." Most of the damage to the soil itself is to a greater or lesser extent repairable, given time, but erosion is a lot more permanent.

The decline in humus levels in soils worldwide is largely, although not solely, a result of modern farming methods. This is a major problem, because humus is the key to soil fertility. It is the product of dead plant and animal remains. This brownish-black organic substance is essential to maintaining a good soil tilth—an open, sponge-like structure—which both maintains good aeration and at the same time has the ability to hold large reserves of water. It contains plant foods in store, for release by micro-organisms when needed by plants. In short, humus is the essential organic ingredient in soil, without which soil life would all but cease to exist.[23] Humus is continuously oxidizing and breaking down, releasing its nutrients, and this is why it has to be replenished with new organic waste on a regular basis. Nitrogen fertilizers speed up the process of oxidization of soil humus—all too common in conventional farming—and when fresh organic matter is not applied, humus levels inevitably decline.

As already hinted at, soil micro-organisms are also a vital part of the story. In soils denuded of humus, micro-organisms decline dramatically. However, the cause is not just the decline in humus levels, but the negative effects of chemical fertilizers and pesticides on soil life of all types, in particular mycorrhizae fungi threads (hyphae), which are vitally important for the absorption of nutrients by many plants as well as the maintenance of plant health.[24] These mycorrhizae hyphae are seriously affected by both chemical fertilizers and fungicides. In some agricultural soils in the United States, the soil is all but dead, and in many intensively farmed areas of the world, soil life is minimal. The soil then becomes just a propping-up medium, and a (not very good) water store for the plants, which get all their nutrients from chemical fertilizers alone. Soils with greatly reduced natural fertility, then, are the legacy the present generation is handing on to the next for them to sort out.

This decline in natural fertility caused by conventional farming methods is exemplified most powerfully by a phenomenon that occurs when farmers convert their farms to organic methods. There is a drop in crop yields during the first few years, then a gradual recovery as the soil life recovers.[25] This phenomenon indicates how denuded the soil life becomes when chemical fertilizers and pesticides are used and good humus levels are not maintained in the soil. Production is at its lowest about two years into the conversion to organic farming, and then there is a fairly steep rise to acceptable production levels over three to six years. However, production continues to improve gradually for about ten to twelve years, and is found to still be gently rising even after fifteen or sixteen years. This is a well-recognized and documented phenomenon, but why does it happen? My wife and I did not know about it when we started organic farming, and assumed that we were doing something wrong. However, after the first three years there were improvements in both yields and the health of the crops. So what had happened?

The first year was not bad, because we were ploughing up grassland to grow our first crops, and this always results in a release of energy, due to improved drainage and aeration. Moreover, the land had had regular applications of fertilizers, and there were residues of those as well. However, the natural life of the soil had been reduced, the soil structure had deteriorated, the land had become acid, and the worm population had been depleted. All this damage had to be undone and true soil fertility rebuilt. In the meantime, the residues of fertilizer and the temporary release of energy from the ploughed-in grass ran out. It needs to be said here that our land is a clay loam on top of shale. When well drained it is slightly on the acid side, has a good humus content and is a good and productive soil. But like any clay soil, when neglected and plastered with mineral fertilizers, the structure deteriorates and it becomes sticky, airless and, as a result, lifeless. Ploughing and cultivating such a soil does aerate it and break it up to a certain extent, but it is not the same as the genuine, more stable crumb structure that only occurs when there is a good humus content, and consequently a high population of micro-organisms, fungal mycorrhizae and worms. The steady improvement in soil structure, good tilth and productivity became increasingly obvious over the following seasons, but it did not happen overnight. It takes time to nurture a soil back into full life, but it did happen.

Pollution

The pollution caused by pesticides and nitrogen fertilizers is one of the main worries that the general public have about conventional farming methods. The main concern is about the pollution of our food (covered more extensively in Chapter 4), but pollution from modern farming takes many other insidious forms. There is the pollution of the country's watercourses, as well as our drinking supplies. The water companies' annual running costs for cleaning up the pesticide and nitrate pollution in the UK's drinking water is around £210–230 million, and the capital investment required to date has been in the order of £947 million to £1.2 billion.[26] Even with the present level of technology, the water companies are unable to extract all the pesticides and nitrates, and many argue that the levels in the drinking water that comes out of our taps are still too high.

Then there is the pollution of the environment and wildlife. The Royal Society for the Protection of Birds (RSPB) is a major supporter of organic farming because of the damaging effects of conventional methods on birds, some of which have declined dramatically over the last twenty-five years, with ten farmland species having declined by 50 per cent or more.[27]

Much of this agricultural pollution ends up in the sea. A huge area of the Gulf of Mexico has been poisoned by the run-off from intensive cattle rearing and maize growing, creating a dead zone 31,000 sq km (12,000 sq miles) in extent.[28]

Intensive Animal Rearing

To many, one of the most insidious aspects of modern conventional agriculture is intensive animal rearing, or to use its pejorative description, 'agribusiness'. Of course agriculture has to pay, and it is a business, but some of the extreme forms of intensive animal rearing have reached obscene levels, where animals are certainly not able "to perform all aspects of their innate behaviour", as required by international organic standards. As Nicolas Lampkin says: "A large section of the public is not prepared to accept the inhumane methods of factory-farming, whether for poultry, pigs or veal, which deny many of the natural behavioural needs of livestock and often create appalling conditions for the workers who have to look after them."[29]

What so often happens in farming with animals, and increasingly in factory-farming conditions, is that the stockmen and women become de-sensitized. There is a gradual corrosion of finer feelings and empathy, and a general dulling of compassion. This then leads to further insensitive behaviour—in fact it is a vicious cycle, which I have seen all too often. The animals suffer in the process, but the real loser is the stockman or woman. This is a deeply unhealthy situation for them to get into.

Dependence on Imports to the Farm

Organic and sustainable farming tries to rely as much as possible on its own resources, particularly when it comes to crop nutrition and animal feed, although the imports of certain types of manure are allowed onto organic farms. Conventional farming, on the other hand, has become increasingly dependent on external imports. This means importing fertilizers, pesticides and often animal feed onto the farm. This is a financial burden, which often reduces profit margins, despite the increase in yields. In industrial countries, farmers are now tied to producing as high yields as possible in order to maintain the increasingly small margins necessary for financial survival.

The problem is usually only solved by increasing farm sizes and raising 'productivity' by reducing staff, whilst at the same time investing in more and larger machinery and becoming increasingly dependent on fertilizers and pesticides. This in turn results in a further dependence on financial loans, which cuts profit margins even more in order to keep up the interest payments. For farmers in the West, this strategy has led to many going out of business and even more barely making a living. Needless to say, for farmers in the Third World this strategy has been even more of a financial disaster.

The Power of the Seed/Pesticide Companies

This dependence on external imports has taken a new twist recently, with the introduction of GM crops. Farmers who have bought into this system have found themselves paying much higher prices for their seed, plus a 'technology fee' for the privilege of growing them. They are not allowed by law to save the seed from their GM crops for re-seeding, because this would break the patent laws that the seed companies have on their seeds, and in the case of herbicide-resistant GM crops, the farmer is contracted to use only the herbicide that the seed company supplies.

In the United States, where GM crops have been in production for longest, a growing number of farmers now rue the day they started down this slippery slope, and are warning farmers in the UK and others in Europe not to take this route.[30] Once again, for many farmers in Third World countries, the extra financial burden of growing GM crops has been the last straw.

Conclusion

It is very easy to make the mistake, when comparing different forms of farming, of assuming that it is a black-and-white issue. This is far from the truth. There is a huge range of farming practice within conventional farming: from almost naturally organic upland extensive sheep farms, to mixed dairy, sheep and cattle farms, on which is also grown pasture, grain and other crops; and at the other end of the scale, to prairie grain farming on a huge industrial scale, and massive intensive animal-rearing factories. However, as long as the present trends continue, the conclusion must be that conventional farming practice is in general not good for the health of the world's soils, for the environment, for wildlife or for humans, and is certainly not good for long-term agricultural sustainability. However there is an alternative. Here are six principles which, if applied to agriculture generally, would transform it:

Six Principles

1. *'We are Part of Nature'* The understanding that we are not separate from Nature is the most important principle of all, and underscores all the rest.

2. *The 'Law of Return'* As in Nature, all waste products should be recycled to feed the next generation. This is the continuous cycle of birth, growth, death and rebirth that has always sustained natural systems.

3. *Sustainability* If the natural system had not been self-sufficient and sustainable it would have ceased to exist aeons ago. Only farming systems that are sustainable have any credibility, and indeed any long-term future. 'Sustainable' means continuing to produce food without a loss of soil fertility throughout time, and 'self-sufficient' means finding all manures and animal foods, as far as possible, from within the farm itself. The term also includes the idea of environmentally sustainable. Destroying soil fertility and the ecology, using huge amounts of fossil fuels to make chemicals and transport them, and having to spend huge amounts of money cleaning up the pollution caused by poisons and fertilizers is not by any stretch of the imagination sustainable.

4. *Biodiversity* This means creating and maintaining an integrated mixed system of farming with as much diversity as possible, in crops, animals, wild plants, trees, water, wildlife and nature reserves. Through biodiversity comes a healthier, self-regulated system.

5. *Healthy Soil* By creating and maintaining a healthy, vibrant, living soil, we will have healthy plants, healthy animals, healthy food and therefore healthy humans.

6. *Respect for Animals* Having decided to take the responsibility of using animals for agricultural purposes, we should treat them as humanely as possible and allow them to live their lives as naturally as possible within the constraints of the farming system.

With the exception of the first principle, these are not listed in order of importance. All of them are equally important, and all are part of the total package, part of the 'holistic enterprise' that is organic farming.

Organic Growth

By using the best of traditional Western farming practices, the best of traditional Eastern practices, the best of improved scientific understanding and the innovative methods that modern organic farmers are continually developing, we can have a very healthy, sustainable farming system that points the way to the future.

The very real threat of global warming and the fact that worldwide oil production will soon peak—if it has not done so already—will mean a decline in supply. As a result, the cost of oil will escalate long before it runs out, so any form of agriculture that can rely less on oil has to be taken seriously. In one Swiss study, organic farming was shown to require between 34 and 53 per cent less energy than conventional agriculture. Several other studies found 60 per cent savings, even when the additional fuel used for weed control was taken into account.[31]

In areas in the world where human and animal power is more commonly used, the reduction in energy use will be even more dramatic. It is becoming clear then that orthodox modern agricultural practice is unsustainable—both in energy terms and in its damaging effects on the environment and human health. The sooner we come to terms with this, the sooner we can invest our energy and creativity in encouraging sustainable forms of farming.

The growth in organic husbandry around the world is an allegory of the changing approach to the way we live on our planet. It is part of a movement that includes the growth of sustainable energy technology, the growth in recycling, and a recognition that we have a great teacher in Nature, of which we are part. Our alienation from Nature has to come to an end. When we treat Nature with disrespect, we are treating ourselves with disrespect; when we pillage Nature, we are pillaging ourselves; when we poison the environment, we poison ourselves (literally); and when we rip off the land, we destroy the very thing that has sustained us and all life on our planet since life began billions of years ago.

Chapter 2

From the Past to the Present

"If I have seen further, it is because I have stood on the shoulders of giants."
—Sir Isaac Newton

The Growth of the Modern Organic Movement

An Amalgam of Ideas

There is a misconception that organic farming is just farming as it was always practised in the West before 1939, without the use of inorganic fertilizers or pesticides, and with only animal manures to maintain fertility.[1] However, modern organic farming is essentially an amalgam of ideas and practices drawn from several sources, largely inspired by four main influences which have resulted in the unique and continuously developing way of farming we see today:

1. The ancient farming methods of such countries as India, China, Korea and Japan;

2. The best of traditional Western farming practices, which stem largely from the agricultural improvements of the eighteenth and nineteenth centuries;

3. The holistic approaches and researches of modern organic pioneers such as Sir Albert Howard, Rudolf Steiner, Jerome Rodale and Lady Eve Balfour;

4. The profound transformation in our scientific understanding of the biological and ecological mechanisms involved.

Take all these influences together, and we have a very different creature from Western farming as it was before 1939.

Farmers for Forty Centuries

Many people think of organic farming and the organic movement as a recent innovation. As Philip Conford noted in his excellent book *The Origins of the Organic Movement*, when he was teaching at a summer school on ecological agriculture held at the Agricultural University of Wageningen, in the Netherlands:

> "Most of the students could trace the organic movement back about ten years, connecting its emergence with the food scares of the 1990s. One or two knew of Rachel Carson's *Silent Spring*, and a middle-aged Icelandic lady was aware that a branch of Rudolf Steiner's 'Biodynamic Movement' had been established in her country by the early 1930s. The general feeling, though, was that organic food production and an ecological approach to agriculture had emerged only since the 1960s." [2]

His book is the most definitive history of the modern organic movement and the pioneering characters that forged it. It is an invaluable work, but I would like to go much further back into history—about 4,000 years or thereabouts. In concentrating solely on the story of the modern organic movement, however pioneering and fascinating that story is, we make the age-old mistake in seeing events purely from the European or Western perspective. Even in the twenty-first century, we still do not question the assumption that runs throughout our education system, that civilization started with the Greeks. It is as though the other great civilizations did not exist, or were of minor importance.

Europe has consistently had a superior attitude towards other, usually older, cultures. Europeans have tended to be blind to existing, well-tried traditional methods of agriculture, whether of Eastern, African, or Native American societies. We have taken our own scientific agricultural paradigms and tried to impose them around the world, without the humility to learn from others.

This tendency continues to the present day, exemplified by the 'Green Revolution' and similar projects, supported by the World Bank, the International Fund for Agricultural Development (IFAD), the International Monetary Fund (IMF), the United Nations Development Programme (UNDP) and the World Health Organization (WHO) amongst others. These programmes invariably involve outside agencies encouraging agricultural practices developed in other cultural conditions, often in the laboratory and usually inappropriate to the new situation, without recognizing or appreciating local methods. Jules Pretty, who has spent many years researching this subject, describes this process in great depth in his book *Regenerating Agriculture*. He gives examples of well established, sustainable agricultural systems that have been completely ignored, and replaced by a 'scientific' model designed elsewhere; for example Native American cultures, which had used sophisticated water-conservation techniques and sustainable agriculture for around 1,500 years, were completely ignored by modern conservationists.[3]

Another example (not in Pretty's book) is that of the European settlers who arrived in the forested area around Mount Kilimanjaro, on the borders of Kenya. They were unaware of the sophisticated and highly productive form of forest farming that existed in the area, which supported a dense population of local Africans. For hundreds of generations the local population had encouraged and planted trees and shrubs useful for food and medicine, reducing the less useful species. In the open areas, they tended crops as well. Here was a three-dimensional, highly productive form of sustainable farming that was not even recognized by the white settlers. Instead, the latter cut down most of the forest to grow sugar cane.

So let us start with a little humility. As Philip Conford notes, it was the humility and wisdom of pioneers like Sir Albert Howard and Dr Robert McCarrison that allowed them to recognize that they had something to learn from other cultures and peoples, rather than something to teach them.

Let us remind ourselves of the Chinese, with some forty centuries of continuous civilization, and with its writings from those times that are still readable today. This civilization invented the iron-chain suspension bridge in the first century AD and the segmental arched bridge in the seventh century, many centuries before the West reinvented them; it was a civilization that was using petroleum and natural gas as fuel by the fourth century BC, and deep drilling for natural gas in the first century BC, as well as using coal and making compressed briquettes of coal and charcoal dust long before we in the West even thought of the idea. In the thirteenth century the wheelbarrow was 'invented' in Europe, but it had been used in China since the first century BC. In my school history lessons, I was taught that it was an Englishman, Jethro Tull, who invented the seed drill around 1700, yet multitubed drills were being used in China in the second century BC. We were told that the production of steel came about with the invention of the Bessemer steel process in 1856, but Chinese were making steel in the second century BC at the latest.[4]

This sophisticated and highly skilled culture, with its arts, crafts, trades, engineering, philosophy, religion, politics and administration, could only have thrived for so long if it had been underpinned by an equally skilled and sustainable form of intensive agriculture, and this was indeed the case.

The remarkable achievements of Oriental agriculture were first described in the West by F. H. King, an American soil scientist, in his book *Farmers For Forty Centuries: Permanent Agriculture in China, Korea and Japan*. In it he describes his observations of the farming practices in these countries during his extensive travels in the early years of the last century, just before chemical fertilizers and pesticides were introduced from the West.

Highly Intensive Production

What King found was remarkable by anyone's standards. Not only had these peoples farmed organically for thousands of years, whilst at the same time maintaining the fertility of their soils, but their production rates were prodigious, to say the least. This was not only sustainable agriculture over forty centuries, but highly intensive agriculture to boot. At that time, the USA was using 1.2 ha (3 acres) of land to feed one American, whilst the Japanese were feeding 3 people off 0.4 ha (1 acre)—in other words, nine times more people were being fed per hectare than in the USA. One of the most productive countries in the West at that period was the Netherlands, yet even that country fed only one person per hectare of cultivated land. China, Korea and Japan together were feeding around five times the number of people per hectare than the USA at the time he wrote.[5] It must also be remembered that whilst chemical fertilizers had not started to be used in the Far East at that period, the USA was using them extensively.

It has to be said that the average weather conditions in these countries are more congenial to intensive production, with longer growing seasons and higher average temperatures along with good rainfall, except in northern China, but this does not fully account for the situation. It is true that populations in these areas were high as a direct result of the good growing conditions, but the increasing pressure from these populations had devastating effects on the environment. The hillsides were denuded of trees and soil erosion grew. As a result a more intensive, but at the same time more sustainable, form of agriculture had to be developed in order to feed the growing number of mouths. The result was the most intensive form of organic agriculture the world has ever seen.

King had left the United States, where there were 8 ha (20 acres) per head, to Japan, which had 0.6 ha (1.5 acres) per head and China which had 1 ha (2.4 acres) per head. He had left a land that had exhausted strong virgin lands and arrived in countries where the land was still fertile after forty centuries. And the reason? Well, necessity is the mother of invention, but there was something else. Ingrained in the minds of everyone in those cultures was the idea of imitating natural processes and learning from Nature.

Recycling

He observed that everything was returned to the land. All the food waste, all the manure, both animal and human, all redundant clothing, and the ashes from fires were returned to the land, either directly or more commonly through extensive compost heaps.

Everything was used; nothing was wasted. Just one example he cited was rape, which was widely grown in China. The young growths were eaten as greens. Oil was extracted from the seeds (as it is today). The seed cake left after extraction was used as fertilizer. The dried stems were then tied in bundles and used as fuel. And finally the ashes from the fuel were used as fertilizer.[6] Chinese farmers knew that water that had percolated through soil had valuable plant foods in it, so they collected the run-off from their terraces by building drainage furrows at their bases, which ran into cement-lined basins. The liquid was used for watering the crops or for preparing liquid fertilizer.[7] Reservoirs were also common, for collecting rain water, for irrigating ridged fields, for breeding fish, for collecting the run-off from the fields, and for providing silt for compost-making. Even the waste water from cleaning pigsties was recycled as liquid fertilizer.

Nonetheless, however careful they were, some of the topsoil washed away every year (especially in heavy monsoon weather) into the ditches and the extensive canals that criss-crossed the land. Every year, therefore, these were dredged for their valuable silt, which was either spread directly onto the land or composted with manure and waste vegetable matter. It was often mixed with clover green manure grown for the purpose and cut from an adjacent field to be composted in huge pits. Around 70 tonnes per acre were spread on various crops such as mulberry plantations, or to fertilize crops of Windsor beans.[8] The extraction of the silt had the added advantage of keeping the ditches and canals open, which was particularly important as they were used, among other things, for transporting human and animal manure as well as green manure to the farms and providing irrigation for the crops.

King also noted that the regular addition of silt had raised the surrounding land over time, thus improving drainage.

Human Sewage Recycling

Whilst in the West we pride ourselves in efficiently disposing of sewage, in the East it was fastidiously collected for use as a fertilizer. Saving sewage from a million people a day provided more than a tonne of phosphorus, more than 2 tonnes of potassium and more than 7 tonnes of nitrogen to feed their hungry soils.[9] And the local authorities, instead of spending taxpayers' money disposing of sewage, were making tidy profits. In 1908, the city authorities of Shanghai sold 78,000 tonnes of human waste to contractors for $31,000 in gold.[10] There was a similar, though less organized, system of middlemen who transported sewage from London's cess-pits to farmers in the eighteenth and early nineteenth centuries, but never on the same scale as the Chinese.

King noted that in Japan, the farmers had screens erected by the side of their fields as public toilets for people to leave precious manure, as a service to both the farmer and the traveller. He also noticed large porcelain containers in gardens for the containment of night-soil, next to which were containers of wood ash and animal manure, each for use in the garden via the compost heap.[11]

Composting and Green Manuring

King discovered in his travels that about 4.5 tonnes of fertilizing material per acre was applied every year in Japan, a large proportion of which was in the form of compost.[12] Herbaceous growth was also cut and carted from the hillsides, along with specially grown crops of clover for composting.[13]

In China he studied and recorded the composting methods employed. Stable and human manure was brought from the towns along the canals, and then mixed with canal mud and composted. In Shantung Province he saw how the local farmers made compost in large pits. The ingredients were soil from the field, saved human and animal manure, coarse straw and other waste vegetation and saved ashes from the fire. These were added to the pit, watered and left to ferment. Shortly before it was ready to use, by which time it had the consistency of mortar, it was spread on the yard to dry and then mixed thoroughly with more fresh soil, brought in from the fields, plus saved wood ashes, and was frequently turned and stirred. Frequently it was spread out and left to dry, and then pulverized by hand or roller, and stored ready for use when it was needed.[14]

In the late 1880s, European scientists demonstrated that leguminous crops provided nitrogen through their root nodules, but Eastern farmers had known for centuries that these crops are essential for the maintenance of soil fertility, whether for composting or green manuring. King noted that *Medicago astragalus* (closely related to alfalfa), was used to eat and later used as a green manure.[15] Chinese clover was often grown with barley, and when the barley was harvested the clover was ploughed into the soil as a green manure for the following cotton crop.[16] Pink clover, *Astragalus sinicus*, was also used as a winter green manure crop, to be followed by such crops as rice.[17] He observed how herbaceous growth was cut from the hillsides and elsewhere, spread between the rows of rice in the paddy fields and then trodden into the mud to feed the rice plants.[18]

Intertillage and multiple cropping

King saw how three crops were grown often together in the same field (intertillage) in ridges and furrows, with one crop in its young stage, another near harvest and the third in mid-growth, drawing most heavily on the soil's resources. These practices, along with irrigation and heavy manuring, produced huge yields.[19] He observed how in orchards an understorey of vegetable crops growing in between the rows of trees was common.[20]

Cotton was often sown in wheat fields as the wheat was ripening, by loosening the soil between the wide rows and then broadcasting the cotton seed amongst the plants. The loosened soil was then flicked over, covering it. This provided two inches of cover and moisture for the cotton to germinate. After the wheat was harvested, the cotton plants were thinned and liquid fertilizer was applied. Thus the cotton's growing season was extended by a whole month, producing better crops and allowing it to be fitted into the cropping year.[21]

Multiple cropping was common. The land was divided into 1.5 m (5 ft) wide ridges, with 30 cm (12 in) furrows between, with the wheat grown on top of the ridges, Windsor beans on the sides, and cotton sown into the wheat as above. Later, after the wheat was harvested, the Windsor beans were harvested while the cotton was still half grown. The sides of the ridges were then composted and cultivated ready for a fourth crop of late autumn greens.[22] In Manchuria, King saw the high ridge cultivation of maize or millet, with alternate rows of soya beans.[23]

He noted that in Japan, millet was often grown between rows of Windsor beans. The beans were picked long before the millet was harvested, and the stems left to dry, with the roots still in the ground to rot to release the nitrogen from their nodules. The stems were then gathered and bundled for fuel.[24]

In Canton, King saw a further example of intertillage. Rice paddies that had previously had two crops of rice in a single year were ridged up in the autumn to enable winter crops of leeks and other vegetables to be grown. Often, when the first crop was half grown, the spaces between the rows were fertilized and then sown with a second crop, which continued to grow after the first had been harvested.[25]

Textiles

At that time, all textiles were made of natural materials. There was, of course, wool from the sheep, which were kept on the hills, but many of the fabrics in these countries were made from crops, such as cotton, hemp, flax and China grass, and of course silk collected from silkworms which fed on the leaves of mulberries grown in plantations.[26]

Lessons for the West

It goes without saying that King was deeply impressed by what he saw and learned from the farmers of the three countries he visited. He was taken by their inventiveness, skills, intelligence and industry. He wrote:

> "Almost every foot of land is made to contribute material for food, fuel or fabric. Everything which can be made edible serves as food for man or domestic animals. Whatever cannot be eaten or worn is used as fuel. The wastes of the body, of fuel and of fabric worn beyond other use are taken back to the field; before doing so they are housed against waste from weather, compounded with intelligence and forethought and patiently laboured with through one, three or even six months, to bring them into the most efficient form to serve as manure for the soil or as feed for the crop." [27]

But it is his fervent belief that the West had much to learn from Eastern farming methods that stands out:

> "In selecting rice as their staple crop; in developing and maintaining their systems of combined irrigation and drainage, notwithstanding they have a large summer rainfall; in their systems of multiple cropping; in their extensive and persistent use of legumes; in their rotations for green manure to maintain the humus of their soils and for composting; and the almost reli-

gious fidelity with which they have returned to their fields every form of waste which can replace plant food removed by the crops, these nations have demonstrated a grasp of the essentials and of the fundamental principles which may well cause Western nations to pause and reflect."[28]

Inspiration

It was his experiences and observations that so inspired the early modern organic pioneers. Here was the ultimate 4,000-year experiment, which showed conclusively that highly intensive organic farming was not only possible, but essential for building and maintaining soil fertility over the centuries. Through many field records kept over this time in China, we know that crop yields have been consistent over this huge time span. Most people in the West today, and I suspect many in the agricultural establishment, are not aware of the productivity of organic, sustainable farming methods over so many centuries.

The Agricultural Revolution

Huge changes took place throughout the eighteenth and nineteenth centuries in European agriculture, including in Britain. The main improvements that occurred during this period largely survived up until the mid-twentieth century. The eighteenth century was a hugely creative and innovative period, which not only kick-started the industrial revolution and made giant leaps in scientific understanding, but also saw major innovations in agriculture.

New European Influences

The main agricultural pioneers in Europe were the Dutch and the French. The Dutch had recognized the value of clovers and had grown them from the mid-sixteenth century onwards, and in the mid-seventeenth century Sir Richard Weston introduced clover from Flanders into England as a field crop. He was also the first to introduce the rotation of crops, using clover as a soil-improving part of the rotation. Clover, with its rhizobia bacteria root associations adds 60–200 kg of nitrogen per hectare per year (50–180 lb per acre), but more than that, it also increases the availability of other nutrients for the crops that follow it.[29]

It is amazing to me that what appear to be established and accepted methods of farming practice in the West are actually recent innovations when seen in their historical context. It was not until the mid-nineteenth century that the underlying principles for effective crop rotations appeared in Britain; the first experiments with rotations in the United States only took place in 1876 in Illinois.[30]

The best-known rotation system was the Norfolk four-course rotation developed by Viscount Townshend on his estate in Norfolk in the 1830s. This involved maintaining fertility in what was mainly a grain-growing area by the judicious combination of cereals, clovers and sheep. The basic rotation, of which there are many variations, was: one year wheat, with the stubble (and weeds) to be grazed by sheep and thus partially manured by their droppings, before spring ploughing, followed by spring-sown oats or barley, undersown with crimson clover. After the harvest of the oats or barley, the clover was then grazed by sheep in the winter and spring, to be grown up for hay in the summer to help feed them through the winter. The clover not only added nitrogen by way of its root nodules, but manure was supplied by the sheep when grazing it. The clover was again grazed at the end of the season before being heavily manured by the stable manure collected from the working horses, which was ploughed in to provide a fertile seed bed in which to grow turnips for the sheep in the winter. Another innovation from the continent, the cultivation of turnips, also provided the opportunity to hoe and weed the ground, making it cleaner for the following crops. The sheep then grazed the turnips *in situ*, thus providing yet more natural manure for the following crop of wheat at the beginning of the new rotation.[31]

The Enclosures

The enclosing of common land by fencing or hedging fields had been going on in fits and starts from the earliest times, with a leap forward in the Tudor period; but it only really took off (and reached its conclusion) between 1709 and 1860, as a result of various Enclosure Acts.[32]

This final and most active period of enclosure made possible major improvements on a far larger scale. The enclosure of land helped to facilitate many of the improvements which still form a large part of both conventional and organic farming practice today. Before the enclosures, local farmers had worked strips of open land around the villages, which restricted farming practice such as the alternation of crops with animals. Animals needed fences, walls or hedges to keep them on the separate plots of land, but the strips were unfenced and open. Instead the livestock were allowed to forage on common land on which the local farmers had grazing rights.

There were, needless to say, many shortcomings to this essentially mediaeval system:

1. The animals had to be monitored continuously so they could be moved onto better grazing and kept off the arable strips.
2. It was difficult to alternate animal stock with crops. Later practice involved the rotation of crops and livestock both in one system.
3. There was little incentive to improve the soil, because the arable strips were rented without security of tenure, and the common land was not owned by the farmers.
4. The arable strips had wasteful gaps between them and were not suitable for the horse-drawn agricultural machinery which was being developed at this time, such as seed drills.

For many lovers of the countryside in the last half of the eighteenth century the fencing in of their beautiful, open countryside and the inevitable clearing of trees was painful to see. The poet John Clare, for example, railed against the destruction of his beloved English landscape:

> Ye fields, ye scenes so dear to Lubin's eye,
> Ye meadow-blooms, ye pasture-flowers, farewell!
> Ye banish'd trees, ye make me deeply sigh—
> Inclosure came, and all your glories fell:
> E'en the old oak that crown'd yon rifled dell,
> Whose age had made it sacred to the view,
> Not long was left his children's fate to tell;
> Where ignorance and wealth their course pursue,
> Each tree must tumble down—'old Lea-Close Oak', adieu!

These lovers of 'the common' and open landscape and heath were as passionately against the planting of hedges and the building of walls and fences, as many are today about their removal!

However, the advantages of the enclosed system of farming were numerous. The new hedges and walls were a secure way of containing animals unattended on specific plots. At last there was a way of rotating both animals and crops at will and in a more organized way. On-and-off grazing is a more efficient form of grass management, as it allows the grass sward to recover between grazings. Resting the pastures helps to control pests such as cattle and sheep intestinal worms. The new walls and hedgerows also provided shelter for the animals in inclement weather; and indeed hedgerows are now recognized as valuable wildlife corridors, as well as food sources and nesting sites for birds.

Ownership of the land involved in the farming enterprise, including grazing land, meant that owners—even those landlords who rented out the land—had a direct incentive for either making improvements or encouraging it amongst their tenants, in terms of the increased profits and an increase in the value of the land. This was also a time of great enthusiasm for new ideas, and many enterprising young landlords were swept up by the modern farming methods coming in from the continent.

It must be remembered that the late eighteenth century and the early nineteenth was a period of innovation, invention and creativity. The Age of Reason had led to unparalleled developments in the sciences, industry and agriculture. The industrial and the agricultural revolutions were but different expressions of the same process, and they fed into each other.

Before 1760 the vast majority of the population lived and worked in the countryside, whereas by the 1830s the majority of the population lived and worked in the new industrial towns. What caused this huge migration of population in so short a time? In previous periods there had been attempts to control and restrict enclosure through Acts of Parliament. What was different about this period was that Acts were now increasingly used to enclose land. If landowners managed to procure the consent of 75–80 per cent of those holding land in the area, an Act of Parliament could be passed to enclose it.[33] This was helped by an increasing enthusiasm for enclosure among members of Parliament, a large number of whom were landowners.

In many areas the commissioners whose job it was to oversee the changes were fair, but inevitably the wealthier landlords managed to increase their holdings and the poor lost out. Smaller-scale farmers were given land to compensate them for the loss of common grazing rights, but they also lost their rights to collect timber, quarry stone, and bracken from the common. Those who had struggled to eke out a hand-to-mouth existence with small-scale subsistence farming were forced, along with the general poor, into the new and growing industrial towns. The lure of regular work in these areas also encouraged the trend, and the countryside emptied.

One thing was certain, however: this new urban population had to be fed! Fortunately, improvements in agriculture increased food production, and at the same time industrial innovations, such as steam-driven traction engines and threshing machines, played their part in agriculture; cultivation became much more efficient, and also allowed more up-to-date machinery to be used in the larger fields.

The Replacement of Best Practice

Unfortunately, conventional farming has to a large extent rejected many of the best innovations and developments of this period, such as crop rotations and the reliance on clovers for building fertility and replaced them with the use of pesticides and chemical fertilizers, seen most noticeably where monocropping is practised.[34] On the other hand, there is still quite a lot of mixed farming left in England, for example in such areas as western Shropshire, where we live. Here farms often combine dairy production, crops grown for animal feed, and cash crops.

On these farms, rotations include a short period when the land is sown to grass for two or three years; but even here arable crops are often grown for many years at a stretch on the more productive plots, simply because it is possible to get away with it by replacing true fertility and controlling the inevitable build-up of diseases with chemicals. If, as often happens, yields start to decrease or pests build up, the land is put down to pasture once again and another plot ploughed up to start the process over again.

Even in the major grain-growing areas, other crops are often grown as 'break crops' to provide a break in the endless cultivation of cereals. These might include beans, peas, linseed or oil-seed rape. Needless to say, none of these approaches is the same as the use of systematic rotations and clovers to maintain fertility and limit disease, which constitute an invaluable part of a truly integrated, mixed farming system.

Influences on Modern Organic Farming

The introduction of clovers, the use of rotations and other improvements, combined with the enclosures, not only gave agricultural practice in Britain a huge boost, leaving a lasting legacy, but also played a substantial role in the way organic farming is practised today. The use of clovers and other legumes in green manures and pastures for both grazing and foraging is seen as essential in maintaining fertility and nitrogen levels in soils on organic farms today, and many organic gardeners regularly use clovers and other legumes for green manuring. Organic farming without rotations of some sort is inconceivable, as different crops take different nutrients out of the soil and have tendencies towards certain diseases.[35] As we shall see in Chapter 3, systematic and well-thought-out rotations not only help to maintain a healthy balance of nutrients in the soil, but also stop the build-up of diseases that affect a particular crop.

As I have said, rotation applies to animals as well as crops. Different animals graze grass differently, concentrating on different parts of the sward, similar to the way that different plants extract different proportions of minerals from the soil. As with crops, the alternation of animals helps to stop the build-up of diseases associated with each animal. Arable crops alternating with temporary grass leys is also an essential part of an integrated farming system, which means resting arable fields every few years by sowing grass mixtures and grazing livestock on them. What the rotation of crops and animals does, therefore, is to parallel the biodiversity seen in Nature; and the greater the diversity of an ecological system, the healthier it is.

Visionaries and Pioneers

We must never forget the enormous influence of the early pioneers of the modern organic movement in structuring organic husbandry as it is practised today. The most important quality they all shared was their holistic vision of Nature and our place in it, and with that understanding, the ability to see how all the parts of an ecological system interrelate. Secondly, pioneers like Sir Albert Howard and Lady Eve Balfour brought a scientific understanding to the processes involved—an understanding of the biology, along with a willingness to carry out scientific trials in the field. A third factor was their willingness to learn from other and older cultures. One need only consider F. H. King's study of traditional Oriental farming, Dr Robert McCarrison's work on the exceptional health of peoples like the Hunzas of northern India, and Sir Albert Howard's studies of Indian and Chinese traditional agricultural methods.

A fourth gift was their ability to be individually creative and inventive, to draw from the best of the past but to add their own unique individual insights into the subject. In this context I am thinking largely of Rudolf Steiner, but also of Sir Albert Howard, Newman Turner and Friend Sykes, who refused to take things for granted and thought for themselves, being prepared to question the accepted wisdoms. And finally they were able to integrate ideas from different sources, which has led to a new approach to farming, containing the best of the old, the best of the new, and the best of their own personal visions.

Let us consider the contributions of a few of the most important pioneers of the modern organic movement. I have already discussed F. H. King and his work in some detail.

Dr Robert McCarrison

It is worth starting our account of the pioneers of the modern organic farming movement with a man who was not a farmer or an agricultural researcher, but a doctor and nutritionist. Dr Robert McCarrison was Director of the Nutritional Research Department in India in the early years of the last century. He was therefore in a unique position to compare the health and diets of various groups in the Indian subcontinent. Some were much healthier and lived longer than others: the Sikhs and Pathans fared better than the Kanarese and Bengalis, but the most impressive of all were the Hunzas. They were renowned for their physical endurance and longevity, and were also very intelligent, witty and civil. These were people who could walk 190 km (120 miles) at a stretch in tough mountain terrain, and who cut two holes in the frozen lakes in winter and swam from one to the other under the ice as a form of sport. McCarrison also observed how they were almost wholly free from disease and lived to a great age. Their neighbours, although warlike, tended to leave them alone because if it came to blows, the Hunzas invariably won. The interesting thing was that these neighbours were also much less healthy although they lived in the same conditions.

To establish whether it was the diets of these different peoples that determined their health levels, McCarrison set up tests in which he fed rats the same diets as the different groups, including a typical British diet of the time. What he found was that the rats' level of health mirrored almost exactly the health of the groups with the same diet; not only that, they suffered from the same ailments as the people whose diets they were eating.

The healthiest rats McCarrison ever raised were, not surprisingly, the ones fed on the Hunza diet. They grew rapidly, were apparently never ill, mated with enthusiasm and had healthy offspring. Throughout their lifetime they were gentle, affectionate and playful. Their diet consisted of unleavened bread made from whole-wheat flour, milk and milk products (butter, curds, buttermilk), pulses (peas, beans, lentils), fresh green leaf vegetables, root vegetables (potatoes, carrots), and fruit, with occasional meat. McCarrison also noted that the healthiest groups were those which ate food that was, for the most part, fresh from its source, little altered by preparation, and complete; and that the agricultural methods used to produce the food involved the complete natural cycle: animal and vegetable waste—soil—plant—food—humans, or animal then humans. At no stage did a chemical or substitution stage intervene.

In contrast, those rats fed on the worst diets not only suffered from the same types of diseases as the people who ate them, but their behaviour and temperament were markedly different. They were vicious and often had to be separated from each other after about six days because they fought and even killed each other. Needless to say, the worst groups were rats fed on a typical British diet of the day, consisting mainly of refined food grown with chemicals.[36]

His book *Nutrition and Health*, along with F. H. King's book, *Farmers for Forty Centuries*, had a huge effect on the early pioneers of the organic movement, who were developing their ideas throughout the first half of the twentieth century. His findings, however, were mostly ignored by the medical establishment. Today the link between diet and health is more widely accepted, but the connection between health and the manner in which the food is grown is less understood, despite a growing body of evidence.

One of the perennial problems has been that medical research has always studied the sick. The Hunzas were a group of exceptionally healthy people who could have been studied further at the time. A golden opportunity was missed, because since then their way of life and diet has changed due to modern communications, influences from the West and the resultant alterations to their farming techniques, which were in the past organic. Fortunately, however, work on the connection between food cultivation and health was continued by Sir Albert Howard, who, among others, studied the Hunzas' agricultural techniques.

Sir Albert Howard

Although organic farming has had a consistent history of at least 4,000 years in certain parts of the world, the organic movement in the West is comparatively young—roughly eighty years. Whenever one delves into the subject of modern organic farming, two giants keep coming into view: Sir Albert Howard and Rudolf Steiner. They were very different men: one was an English mycologist and agricultural lecturer, the other an Austrian philosopher, scientist and artist, and founder of the anthroposophical science of the spirit. Although Howard can be said to have had a more English pragmatic nature and Steiner a more spiritual one, they both shared a holistic vision of Nature and our place in it, despite approaching it from very different perspectives. Both were also deeply practical and developed procedures and techniques of farming that are largely followed today by organic farmers.

Howard started as he meant to go on, gaining a first-class degree with distinction in chemistry at the Royal College of Science in 1893, then gaining a first in natural sciences at Cambridge in 1897, and in 1898 was second in all England in the Cambridge Agricultural Diploma. He started his career as a mycologist, studying fungal diseases for the Department of Agriculture in Barbados. Many of his generation went out to work in the numerous British colonies with the idea of providing scientific expertise and advice to the local farmers. Most had such a deep-seated belief in the superiority of their approach to agricultural research, theory and practice that they were often unable to see value in the existing local methods. It was their job, as they saw it, to civilize the rest of the world and provide the different races with the latest in agricultural knowledge and technology.

What distinguishes Howard was that he was wise and open-minded enough, even from his earliest days in the West Indies, to learn from the native farmers. It was whilst on a tour of the Windward and Leeward Islands, where he was advising the locals on how to grow a whole host of plants such as groundnuts, bananas and citrus fruits, that he found he was beginning to learn more from the people that actually worked on the land than from those who beavered away in secluded laboratories. As he said of this period in his life: "I was an investigator of plant diseases, but I had myself no crops on which I could try out the remedies I advocated. It was borne in on me that there was a wide chasm between science in the laboratory and practice in the field."[38]

What he really wanted was a post where he could combine both theory and practice, and his chance came in 1905 when he was offered the job of Imperial Botanist to the Government of India, at the Agricultural Research Station of Bengal at Pusa. By this time he had already started to formulate the ideas that he was to develop later in his career. He was becoming increasingly convinced that a plant's ability to resist disease was determined by its health, which in turn depended on the health of the soil. With that in mind he set out to grow such healthy plants that they would suffer from so little disease he would not have to use poison sprays.

Whilst in Pusa he made a careful study of traditional Indian agricultural practices, regarding the native farmers as his teachers. At that time the Indians did not use pesticides or artificial fertilizers, and manured their land carefully with animal and vegetable waste. He copied their methods, and by 1919 he was so successful in growing healthy crops that he was able to say of himself that he had learnt "how to grow healthy crops, practically free from disease, without the slightest help from mycologists, entomologists, bacteriologists, agricultural chemists, statisticians, clearing houses of information, artificial manures, spraying machines, insecticides, fungicides, germicides, and all the other expensive paraphernalia of the modern experimental station."[39]

He also discovered that the work-oxen on the farm, which were fed solely on crops grown on his fertile land, never succumbed to the diseases that so often devastated the other cattle in the area around Pusa, such as foot-and-mouth, rinderpest and septicaemia.

At the same time he had been studying the age-old practices of China and had decided to implement these methods at Pusa. Unfortunately, at that time the research station was becoming over-specialized and divided into a series of water-tight compartments: plant breeding, mycology, entomology, bacteriology, agricultural chemistry and practical agriculture. It was his nature, on the other hand, to see things holistically, to understand the interconnectedness of all the parts. As a result he found himself hindered and limited by the departmentalized and specialized structure, and more and more estranged from his colleagues, who were travelling a different path towards some 'brave new scientific world' to which he had no intention of going.

His only hope was to collect funds laboriously so that he could set up his own institute, which he could run his own way. Eventually, with the help of the central Indian authorities, he achieved his aim and set up the Institute of Plant Industry at Indore, 1,300 km (800 miles) north of Bombay, in a cotton-growing area. Here he developed his 'Indore process' of compost production, based on what he had learned from the Chinese. Within a few years he was producing cotton with yields three times those of local farms and with a minimum of disease.

As a scientist he wanted to understand the processes involved in good composting so as to produce the best quality product without any loss of nutrients. What he discovered, however, was something far more important: that he could increase the nitrogen content of a compost heap if it was made in the right way. Through his research he discovered the principles that are still used today by organic farmers the world over.

His genius was to combine an integrated and holistic vision with a detailed scientific understanding of the processes involved in both natural ecosystems and agriculture. It is also to his great credit that he was able to maintain his integrity and continue to follow his ideas, despite the fact that his scientific colleagues were travelling in a different direction altogether. It takes great courage to risk your career following unfashionable ideas. Although many farmers around the world liked those ideas enough to implement them, his scientific colleagues ridiculed him.

When he came home to England at the end of 1935 he was invited by the students of the School of Agriculture at the University of Cambridge to address them on 'The Manufacture of Humus by the Indore Method'. Most of the staff also got to hear about the lecture and turned up. Nearly all of them, from chemists and pathologists to plant breeders heatedly disputed his remarks, although the student body seemed enthusiastic. As Howard recalled, he was "vastly amused at finding their teachers on the defensive and vainly endeavouring to bolster up the tottering pillars supporting their temple."

> "Here again I was amazed by the limited knowledge and the experience of the world's agriculturists disclosed by this debate. I felt I was dealing with beginners and that some of the arguments put forward could almost be described as the impertinences of ignorance. It was obvious from this meeting that little or no support for organic farming would be obtained from the agricultural colleges and research institutes of Great Britain."[40]

This shows how far ahead he and other early pioneers of the organic movement were; it is only in the last decade or so that some agricultural colleges have started to run courses on the subject. It is my experience that academic institutions, with their professional vested interests, are the last places to absorb radical new ideas, and not only in agriculture.

The first book Howard wrote was *The Waste Products of Agriculture: their Utilization as Humus*. At the end of his career he produced his magnum opus, *An Agricultural Testament*, which summarizes his ideas and life's work. It is still in print, and can be purchased from the Soil Association.

Rudolf Steiner

Rudolf Steiner was a seer who had studied the philosophy of Goethe, and edited an edition of the latter's scientific writings in the 1880s. Between 1889 and 1896 he worked on the standard edition of Goethe's works at Weimar. During this period he wrote *Truth and Science* for his PhD. His best-known work, *Die Philosophie der Freiheit* (The Philosophy of Freedom) was published in 1894. Around this time he started to develop a facility for 'spiritual perception independent of the senses'. As a result of these personal experiences he developed a philosophy and system of gaining knowledge which he named Anthroposophy, and in 1912 founded the Anthroposophical Society.

This enlightened man left a worldwide movement and many practical applications of his ideas. The list of establishments based on his work is impressive: the Waldorf School movement, based on his ideas on education and the development of children's minds; homes and schools for handicapped and maladjusted children; a therapeutic movement with a central clinic at Arlesheim in Switzerland; centres for scientific and mathematical research; Eurythmy, the art of movement and speech to music; schools of drama, speech, painting, sculpture and architecture; and last but not least, the biodynamic system of organic farming.

In 1924, one year before his death, Steiner was asked by some farmers to give a series of lectures, which became the basis for biodynamic farming practice. In essence, Steiner saw the Earth as a living organism, with the plant life on its surface as its skin, the interface between the Cosmos and the Earth, with their roots in the soil and their heads in the air. He saw the farm also as a self-interacting organism with the farmer as an integral part. In his philosophy the diverse system of different crops, trees, water, hedges, farm animals, wildlife, living soil and farmer are but parts of a unified whole. The farmer is aware of the rhythms of the Earth and the Cosmos and their effects on the farm, with certain forces at work in the morning, making it more appropriate for certain farming activities, and different forces at work in the afternoon, which are more appropriate for others. The farmer is also aware of cosmic rhythms which affect the Earth, so plants, harvests and performs other farming activities at certain times of the year wherever possible. These effects have been studied extensively over the years and show improved yields and keeping qualities in the crops grown.[41]

Steiner also prescribed various preparations to be applied to the soil and the crops, in addition to compost. These preparations can be seen as analogous to homeopathic medicines, being both 'enlivened' and used in small proportions, although his agricultural doses are not nearly as small as those in homeopathy. The first preparation is made from cow manure, its purposes being to promote root activity, to stimulate the soil's micro-life, to regulate the lime and nitrogen content of the soil, and to help the provision of trace elements. The second is made from very finely ground quartz or feldspar, and is used as a foliar spray to enhance the light and warmth which encourage plants just at the time when they are starting to develop that part of the plant which is to be harvested. The compost preparations, made from specially prepared herbs, are added to newly made compost heaps. They act more like regulators rather than activators, having a balancing and harmonious effect.

Biodynamic farming is practised worldwide and as a result its standards are consistent the world over. All the products produced on biodynamic farms are grown to the same rigorous standards, whatever the country of origin, and are sold under the name 'Demeter'. From my own research, when comparing the yields of organic farms, it appears that those on biodynamic farms are consistently higher. However unusual the approach may seem at first, it definitely produces results. After farming organically for about ten years, I became interested in biodynamic farming and practised its methods for three or more years. This is not really long enough for a serious study, but I must say even in that short period of time I was impressed with the results. Of course it was not a comparative study, but I did find on a personal level that it added another dimension to the way I saw farming and my place in it, which gave me added satisfaction and understanding in my daily farming practice, apart from the practical benefits.

In 1990 there were only fifteen biodynamic farms out of 665 organic farms in the UK (2.5 per cent), which may be due to the Britons' pragmatic, and dare I say it, more cynical approach; but on the Continent, especially in Germany and Steiner's home country Austria, it is by far the dominant method. There are also large groups in both North and South America, New Zealand and Australia.

Lady Eve Balfour
The next step was for practising farmers to take up the challenge. The first notable farmer to follow Howard's methods in the UK was Lady Eve Balfour, who just before the Second World War started using his techniques on her farm at Haughley in Suffolk. She had suffered from a number of conditions, including bad head colds every winter for years and bad bouts of rheumatism since childhood, especially in prolonged periods of cold, wet weather. She began eating the organic produce from her farm, including her own bread made from her now organic flour. During the winter following her change in diet, she was both free from colds and no longer bothered with rheumatic pains.

After a prolonged period of research and detective work, as well as interviews with health specialists who were convinced of the soundness of Howard's and McCarrison's ideas, she wrote her now famous book *The Living Soil*, which was first published in 1943. Its whole emphasis was on the relationship between diet and health, and she contended that the health of the soil has a direct relationship with the health of those who eat the food grown on it. The book was published during the war, which was probably not the best time, considering that every scrap of land was being used to produce as much food as possible by whatever means necessary. Add to that the fact that the nation's attention was elsewhere, and it is surprising that the book had any effect.

Lady Eve was aware that there was much hostility to organic ideas from the scientific community, so she decided, with the help of others, to start a systematic research programme at Haughley. In 1976 *The Living Soil and the Haughley Experiment* was published. The first part was an edited version of the original book to meet the increasing demand for its republication. The second part described the Haughley experiment and the details of her findings over twenty-five years. Among its findings were that the levels of available minerals fluctuate according to the season, and that the average humus level is higher in organic than in mixed conventional farming, and as a result organically grown crops utilize their soil environment more efficiently.

For forty years Lady Eve collected and collated evidence to show that the health of the soil, plants, animals and people are indivisible. She was a pioneer researcher into the nutrition cycle. Nowadays we take for granted the research being carried on around the world into organic farming, but she was the originator.

In the UK her work is being continued at the Elm Farm Research Centre, which was a founding member of the European Consortium for Organic Plant Breeding (ECO-PB) and members of the International Research Association for Organic Food Quality and Health. Elm Farm not only carries on experiments, but also provides an Organic Advisory Service using experienced organic practitioners to provide direct advice to farmers and growers.

Lady Eve was of course the founder, along with Friend Sykes and others, of the Soil Association in 1946, one of her greatest legacies. It has been instrumental in promoting organic farming in Britain over the years, organizing research and providing information on the links between the way food is produced and human health, as well as the effect on the environment. It is best known for its work developing high standards for organic food production and its symbol scheme, with its certification and labelling of organic food in the UK.

Friend Sykes [42]

After the war, others followed in Lady Eve Balfour's footsteps, including Friend Sykes, an English farmer, breeder of thoroughbred horses and agricultural adviser. He bought a derelict 300 ha (750 acre) farm in Wiltshire with the idea of farming it in accordance with Howard's ideas. He had come to the conclusion, through his experience over the years as an agricultural adviser, that restricting the number of crops grown and farming with only one type of animal inevitably led to a weakening of stock and plants by disease. He was convinced that with a combination of good husbandry and mixed farming, outbreaks of disease could be eliminated.

He was ahead of his time in understanding the principles of ecology before the term became a household word, and he could see very clearly how pests and diseases would become resistant to the onslaught of chemicals like DDT. In 1950, his book *Food, Farming and the Future* was published. In it he wrote:

> "The first thing that Nature does when she has been treated with poison is to battle against it and try to breed a resistant strain of the form of life that is being attacked. If the chemist persists in his poisonous methods, he often has to invent more and stronger poisons to deal with the resistance that Nature sets up against him. In this way, a vicious cycle is created. For, as a result of the conflict, pests of a harder nature and poisons still more powerful are evolved; and who is to say, in this protracted struggle, man himself may not ultimately be involved and overwhelmed?" [43]

His ideas have been entirely borne out since then. We now know that, despite a tenfold increase in the use of insecticides since the First World War, the loss of food and fibre crops to insects has risen from 7 to 13 per cent, and that at least 520 species of insects and mites, 150 plant diseases and 113 weeds have become resistant to the pesticides meant to control them.

Sykes had a laboratory analysis done of the soil on one of his fields, which showed severe deficiencies of lime, phosphate and potash. They recommended that he used lime and artificial fertilizers to correct the shortages. He ignored the advice, believing that soil had a latent fertility that could be encouraged just by tilling it. He set about ploughing and harrowing the field, which he then sowed with oats. Much to the amazement of his neighbours, the crop yielded 4.3 tonnes per hectare (1.75 tons per acre). Again without adding anything, he tilled the land and sowed a crop of wheat. The yield was similar. He sent a new soil sample to the laboratory and it was found that the acidity had rectified itself and potassium levels had been restored, but there was still a deficiency of phosphorus. The laboratory recommended a heavy dressing of phosphates. Again he ignored their advice and bought a sub-soiler, a tool that breaks up and aerates compacted subsoil. He sub-soiled the field and prepared it for another crop, again for wheat. The yield was even better than before, and a further analysis of the soil showed no deficiencies at all!

The only explanation I can find for this is that the phosphate and potassium were present in the soil in unavailable form, and that they became available as the biological activity in the soil was activated as a result of the increased aeration due to ploughing and sub-soiling. As Nicolas Lampkin explains in his book *Organic Farming*, "A range of standard tests exists to try to estimate the available fractions of crucial nutrients like phosphate and potash, but often, particularly with phosphate, the tests give no indication of total reserves or the proportion of total reserves which are available to the crop." [44]

After this experiment Sykes went on to apply compost to his land on a regular basis, because this strategy would not have continued to yield results for very long. Nonetheless it was a valuable and revealing experiment.

Most farmers would have accepted the experts' recommendations. The land in the area was considered to be poor, but like many pioneers and free thinkers he was prepared to experiment to find out what would happen, based on his innate belief in Nature and the soil's ability to correct itself, given the chance. He went on to write an essay entitled 'Farming for Profit with Organic Manures as the Sole Medium of Refertilization'. In it he describes how he not only made money from his farm, but he had created crops and animals that were disease-free, and he had managed to do it without the use of poisonous sprays, which mirrors almost exactly the findings of Sir Albert Howard and Newman Turner.

J. I. Rodale [45]

At about this time, things were also stirring in the United States. Rodale was a Renaissance man with boundless energy. He was an accountant, inventor, electronics manufacturer, creator, entrepreneur, and went on to become an organic farmer, author, publisher and playwright. In the 1940s he edited a health magazine in Pennsylvania, and it was when he was there that he came across the work of Sir Albert Howard. He was stunned by Howard's contention that the way food was grown had a direct effect on its nutritional quality. As he wrote, "This theory had not found its way into the articles of any of the health magazines I was reading. To physicians and nutrition specialists, carrots were carrots were carrots."[46]

He was so inspired by Howard's ideas that he organized the publishing of *An Agricultural Testament* in the United States, bought a farm of his own and launched his hugely successful magazine, *Organic Gardening and Farming*, along with its sister publication, *Prevention*. With both, his purpose was to educate the American public about the living soil; how it was clean and thriving with life, and how by feeding the millions of micro-organisms, in the soil, healthy crops could be grown, which in turn would be healthier for those who ate them; in other words how there was a direct connection between organically grown food and health.

Louis Bromfield [47]

Louis Bromfield rivalled Rodale as the leading exponent of organic ideas in the US in the 1940s and 50s, and was very influential in the development of the organic movement there, although he criticized those in the organic movement he saw as 'fanatics', and in his farming practices continued to use chemical fertilizers judiciously. Nonetheless, he aimed most of his fire at industrial agribusiness and the growing corporate food industry in the States.

In the 1930s he travelled to India several times, and on one such visit he went to Sir Albert Howard's experimental farm at Indore and was very impressed by the results of Howard's experiments. At the end of the second world war he bought Malabar Farm, and became one of the best known farmers in the United States, largely due to his books: *Pleasant Valley*, *Malabar Farm*, *Out of the Earth* and *From my Experience*.

He was, like Howard, a religious and spiritual man, but as he said of himself, he was a pragmatic man because of his Protestant upbringing and Anglo-Saxon blood, and his faith was based upon the earth itself and demanded concrete results. He believed that agriculture was the keystone of a nation's economic structure, and that the wealth, welfare, prosperity and even a nation's freedom were based upon the soil. He believed that it was vital to build up the fertility of the soil, seeing this as the source of any nation's power. He described true agriculture as a process of understanding 'natural and universal laws' and contrasted this approach with the exploitative agriculture practised in the United States.

Like Howard, he was dismissive of experimental agricultural scientists who spent their time in laboratories, whom he termed 'laboratory hermits'. He wanted to see much more contact between agricultural colleges and farmers practising in the real world.

Edward Faulkner [48]

Edward Faulkner's claim to fame was his hugely influential book *Ploughman's Folly*, published in 1943. In it he puts the case for doing away with the plough altogether. This may sound like a complete contradiction of Friend Sykes' ideas, but it is not. Sykes had compacted soil which, when opened up, reactivated the life within it, making available the plant nutrients. Edward Faulkner, on the other hand, began to realize that organic matter tends to putrefy when buried too deep, that most soil organisms involved in converting organic matter into plant foods live in the top few centimetres. A lot of the goodness is therefore wasted by burying it. However, as Newman Turner discovered, one can still aerate the soil by sub-soiling, without burying the compost and other organic matter by ploughing.

Instead of burying organic matter with the mouldboard plough, it was better to leave it on the surface, or at the most to disc it into the topsoil's upper layer. This would produce the ideal tilth, both granular and water-retentive—in fact the perfect conditions for growing healthy crops. Faulkner also argued, as many have before when questioning accepted practices, that just because mouldboard ploughing had been consistently used for at least 200 years, it did not mean it was the best way of cultivating the soil.

Edward Faulkner forms a neat link between Louis Bromfield and Newman Turner, because it was Bromfield who helped Faulkner's book to prominence, by devoting a whole chapter to his ideas in his book *Pleasant Valley*, and it was Newman Turner who so enthusiastically took up his ideas and practised no ploughing on his farm in Somerset in the 1940s and 1950s.

Newman Turner [49]

Newman Turner's books *Fertility Farming*, *Herdsmanship* and *Fertility Pastures* were our main inspiration before we started farming, and a practical guide during the early years when we were converting our farm to organic practice in the early 1970s. Turner had a no-nonsense confidence in his own beliefs born of his undoubted success in turning what had been an ailing and disease-ridden farm into a thriving, healthy and profitable one. He wrote:

> "It is not in increased yields, or in costs, that I measure the success of this organic fertility farming, though these things are important in times of economic stress. It is the health of all living things on the farm that proclaims nature's answer to our problems. From a herd riddled with abortion and tuberculosis, in which eight years ago few calves were born to full time, and those few that did reach due date were dead, I can now walk around sheds full of healthy calves, and cows formerly sterile, now heavy in calf or in milk."

And later:

> "Such restoration of a dead farm is an achievement worth any man's efforts, and success within the reach of any farmer who will turn back to fertility farming, and eschew the 'get-rich-quick' methods of commercialised science, which are in fact a snare."[50]

The seven-course rotation we used on our farm was directly based on Turner's eight-course rotation, adapted to our conditions. Of particular value was his use of deep-rooting herbs in the four-year grass leys, which both helped to break up and aerate our heavy soil, and also brought valuable minerals and nutrients back up to the surface, for the health of both the animals and the land. We also followed his method of making hay on tripods, copied from the Scots, which was ideal for our small farm, and ensured that we always had good quality hay despite the vagaries of the English summer.

Turner was not one to turn his back on the latest ideas and innovations either, where they fitted into the organic scheme of things. He took to silage in a big way at a time when it was still fairly new, despite the fact that in those days there was no modern machinery to make it comparatively easily. It all had to be lugged, loose and unchopped, and when it was ready for use, it had to be cut by hand from the clamp by a hay knife, a sweaty job for the best of men. He also experimented with lucerne on his Somerset farm.

Like Sykes, he frequently refused to follow 'expert' advice, following his own intuition and common sense, often with surprising results, and the dedication in the front piece of his book says it all: "To my mother, who taught me to think for myself."

Lawrence D. Hills

Lawrence Hills, with his wife Cherry, created a unique organic gardening organization, the Henry Doubleday Research Association (HDRA) in the mid-1950s, which has inspired many generations of Britons to garden organically. He was a skilled nurseryman and gardener, and made an enormous contribution to the organic movement as a whole. In 1971, when my wife and I joined HDRA, there were only are few hundred members, and the association was based at Bocking near Braintree in Essex, with only 1.2 ha (3 acres) of ground, as I remember it. In the early 1980s they moved to their new headquarters at Ryton, near Coventry, and have now over 30,000 members. It has grown into one of Europe's, and indeed the world's, largest centres for organic horticulture, due largely to Lawrence Hills' energy, vision and endless infectious enthusiasm.

Hills was unique. His enthusiasm for his subject, indeed for life in general, knew no bounds. He was classically educated, and when he stayed with us for two nights in the late 1970s, researching a book on organic smallholdings, his conversation was often interspersed with quotes from Xenophon or Sophocles. He was truly a Renaissance man, and a great inspiration.

HDRA was our guiding light in the early days, because members were encouraged to take part in experiments each year and to send in the results at the end of the season. There was no expert telling one how to grow organically, just encouragement to be directly and personally involved in a movement which was constantly learning new techniques and testing them out in the field. This atmosphere of continuous experimentation, dynamic development and open-mindedness that Hills generated at HDRA—characterized by the organic movement to this day—was also characteristic of Sir Albert Howard.

Scientific Insights

The biological and ecological mechanisms involved in Nature, and in the soil and ecology of a healthy farm, have a huge influence on organic husbandry. Anything that gives us a greater insight into the functioning of Nature can only help us to improve our practice of agriculture. The problem in the past has been the specialized vision of chemists, with their limited understanding of plant nutrition.

It was Howard, with his scientific background but integrated vision who tried to understand the scientific reasons for the success of the Oriental system. He knew the ancient methods worked, but understanding how they worked led to a more systematic approach to organic practice.

One example that springs to mind is the scientific understanding of the importance of the role of mycorrhizae fungi in the nutrition and health of plants. Once recognized, it encouraged Howard to use farming practices to create the ideal conditions for their maximum development, and to try to avoid methods that interfered with their production.

Another example was his study of the methods of the Indians and Chinese in creating humus through composting. Through his scientific understanding he was able to gain an insight into the processes involved and to develop a formula for producing compost of consistent value.

Another scientist whose career covered roughly the same period as Howard's was Sir John Russell. He worked at Rothamsted Experimental Agricultural Station as a soil scientist, and was one of its Directors. Russell wrote his now famous book, *Soil Conditions and Plant Growth*, known by many generations of agricultural students. It was first published in 1912, ran to ten editions, and was last published in 1973.

Places like Rothamsted and people like Sir John Russell were dismissed by the early pioneers of the organic movement, including Howard, as classic examples of the 'NPK' mentality—the belief that all that is required for the growing of crops is a selection of water-soluble chemicals: nitrogen, phosphorus and potassium. Howard disdainfully called Rothamsted "the Mecca of the orthodox". A lot of his criticisms I would agree with, but what Russell and his contemporaries did so well was to study the complex biological processes in a living soil—precisely the sort of research that has vindicated the ideas of organic practitioners and their supporters, as well as adding another level of practical understanding.

Another level of scientific research is carried out directly into organic methods through studies done in the field. Much research is being carried out in countries around the world, like that at the Elm Farm Research Centre.[51] There is also a twenty-one-year comparison which was begun in 1978 by the Swiss Federal Research Station for Agroecology and Agriculture and the Research Institute of Organic Agriculture. The team, led by Paul Mader, recently published their findings in the magazine *Science*.

They used four comparison plots in the trial:

A standard organic plot;
A biodynamic plot;
A conventional plot using both farmyard manure and mineral fertilizers;
A conventional plot using only mineral fertilizers.

As had been found in other trials over the years, the yields from the organic plots were 20 per cent lower on average than those from the conventional plots, but this was more than offset by the vastly improved efficiencies in the organic systems. There was a reduction in energy input, including fertilizers, of between 34 per cent and 53 per cent, and a reduction of 95 per cent in pesticide use (and those pesticides that were used, were only those approved for organic use). In other words, the organic systems used resources more efficiently, producing more for each unit of energy and other inputs than they consumed.

The scientists also found a much larger and more diverse community of organisms. There were much larger populations of soil microbes and mycorrhizae (root-colonizing fungi), both of which govern the nutrient cycle. The soil microbes help to break down organic matter to release plant nutrients, and the mycorrhizae, as Sir Albert Howard knew so well, are essential in a healthy soil in helping the plants to absorb nutrients. The researchers said that mycorrhizae fungi were at least partly responsible for the improvements observed in the soil structure in the organic plots. The other type of organism that is fundamental in maintaining and creating good soil structure is, of course, the earthworm, which the scientists found was more abundant on the organic plots. Another benefit of the larger populations of soil microbes, mycorrhizae and earthworms was the improved and more efficient breakdown, and resultant release, of essential plant nutrients.

Insects were not only more abundant but more diverse on the organic plots, with much larger populations of beneficial insects such as pest-eating spiders and beetles.

The researchers concluded their report by saying,

"The organic systems show efficient resource utilization and enhanced floral and faunal diversity, features typical of mature systems. We conclude that organically manured, legume-based crop rotations utilizing organic fertilizers from the farm itself are a realistic alternative to conventional farming systems."[52]

The Best of East and West, Old and New

The lesson of this chapter must be to take the best practices from the past, from our own and other cultures, and combine them with the best of the new understanding that modern science has to offer, as well as all the pioneering work and insights of the founders of the modern organic movement. We should not forget, however, that the experiments and growth in both practical knowledge and understanding are continuing. Around the world, farmers are applying the basic principles outlined in this book to their own geological, social and climatic conditions, sometimes combining them with traditional local techniques, often with exciting results. Some of these fascinating developments are discussed in the final chapter of this book.

Chapter 3

The Living Soil

"Little or no consideration is paid in the literature of agriculture to the means by which Nature manages land . . . Nevertheless, these natural methods of soil management must form the basis of all our studies of soil fertility."
—Sir Albert Howard [1]

Soil Ecology

The world of life in the soil is fascinating in itself, but it is also important because knowledge of the living soil and its health is essential for our understanding of organic husbandry. One of the most important revolutions in modern organic farming has been the growth of knowledge of soil science and biology over the last hundred years or more. If you are attempting to work with Nature as a practising organic farmer, the most powerful tool at your disposal is knowledge.

Of course there is no substitute for knowledge gained by 'doing' and accumulated experience; or as my grandfather, a professional horticulturist, used to say, "You can't learn about growing plants from books." Nevertheless, the accumulated evidence of the complexity of life in the soil over the last century has had a profound influence on organic practice.

The more that scientific researchers study the soil, the more amazed they are at the sheer diversity of life within it. Soil is not just a heap of inert rock particles ground down by ice and water, wind and frost over millions of years. It includes that as its basis, but it is also a vibrant living ecosystem comprising a huge range of micro-organisms, including different types of bacteria, actinomycetes, fungi and algae, microflora, larger plants growing in the soil, protozoa, worms of many different types, arthropods and a whole range of animals. Most of these creatures live on a variety of organic residues made up of decaying plant and animal material from all the organisms that have lived and died within and on top of the soil—in other words, that most miraculous of materials, humus. Add to this mix life-giving air and water, and you have a wonderfully complex system that is in a state of continuous and dynamic interaction and change—a vast living community, exchanging, converting, eating, killing, reproducing, absorbing, transforming, unlocking, dying, releasing, oxidizing, trading, co-operating, competing, and forming mutual alliances. The vast majority of life on the Earth lives within the top few centimetres of the soil and the top layers of the sea. This thin living crust supports all the life on earth; all the plants, forests, animals, birds and insects as well as us humans, and we interfere with and destroy this life-support system at our peril.

Humus

Before we consider the organisms that inhabit the soil, we have to understand the nature of the home in which they live, and above all humus. Humus is the product of dead plants and animals in different stages of decomposition. Some has reached the stage where it is highly resistant to further decomposition and is in a relatively stable condition. Some of it is still in the process of breaking down through oxidation and reduction, largely caused by the micro-organisms that live on it. These micro-organisms not only excrete waste material, but when they die their bodies also become plant food.[2] The carbon content of humus is usually 55–58 per cent, and the nitrogen content 3–6 per cent. This is one of the most interesting things about humus: whatever proportion of carbon and nitrogen in the original plant or animal remains, when it has decomposed it will end up in a stable proportion of approximately 10 to 1; not surprisingly, this is the ideal balance for soil and plant life.[3]

To simplify things, young green plants, animal remains and manure have a very high proportion of protein, which largely consists of nitrogen. On the other hand, older plants and woody material have a much higher proportion of cellulose and lignin, which are largely composed of carbon. When animal manure or other material rich in nitrogen breaks down, the excess nitrogen is released as ammonia, and the resulting humus ends up with a carbon-nitrogen ratio of approximately 10 to 1. When older plant material such as straw breaks down, the bacteria helping the process use nitrogen from the atmosphere to build their own bodies, and this enables them to decompose the high-carbon plant material—again ending up with a carbon to nitrogen ratio of around 10 to 1. The exception is when high cellulose and woody material begins to decompose in wet, airless and acid conditions, such as peat bogs. Then the material remains unrotted and builds up to great depths over time.

Humus is essential because it is like a storehouse or larder of food and energy. As Sir Albert Howard so eloquently puts it in *An Agricultural Testament*:

> "Humus is not in a static, but rather in a dynamic, condition, since it is constantly formed from plant and animal residues and is continuously decomposed further by micro-organisms. Humus serves as a source of energy for the development of various groups of micro-organisms, and during decomposition gives off a continuous stream of carbon dioxide and ammonia. Humus is characterized by a high capacity of base exchange, of combining with other soil constituents, of absorbing water, and of swelling. . . . To this list of properties must be added the role of humus as a cement in creating and maintaining the compound soil particles so important in the maintenance of tilth."[4]

Humus is therefore also invaluable for its physical properties: its ability to absorb and hold water, and as a vital constituent of soil structure by both opening the soil and attracting clay particles into the formation of soil crumbs—an understanding so often missing from the refined perspective of a chemist.[5]

Soil Micro-organisms

A soil rich in humus is home to an almost unbelievably diverse population of bacteria, inhabiting many different niches and varying habitats, even within a small area. Populations vary hugely, depending on different soil types or soil temperature, and can even vary over a twenty-four-hour period. However, a good average figure for a normal healthy soil is about 600 million bacteria per gram of soil![6] A healthy soil consists of soil crumbs made up of smaller particles that have collected together, similar to a coarse pastry mix before adding the water. This makes for a very varied environment, from films and pockets of water, to air spaces comparatively rich in oxygen, to areas at the centre of larger soil crumbs where there is little or no oxygen. There are enough types of bacteria to utilize every type of condition: some specialize in one type of environment, yet others are able to adapt their metabolism so they can survive in both oxygen-rich environments and those where oxygen is scarce. The populations of some types of bacteria remain fairly constant, whilst others increase rapidly to suit the changing circumstances, only to die back and hibernate in the form of spores waiting for the next time the conditions are ideal, when they can come to life again. The bacteria have many different roles. Some specialist ones busy themselves breaking down organic proteins into nitrites; others convert nitrites into ammonium; while others convert ammonium into nitrates, thus helping to release and make available vital nitrogen for the growth of plants in this fascinating chain.[7]

Here the story becomes even more intriguing, because there are certain specialist bacterial populations which are much denser within the root zones (rhizosphere) of plants because they benefit from a symbiotic relationship with them. In natural conditions, plants would starve without bacteria and mycorrhizae fungi. They do not have a digestive system, so the micro-organisms do the job for them, in return for food that the plants provide. Plant roots exude a mixture of substances known as mucigel, which contains simple sugars, protein, carbohydrate, growth factors, organic acids and enzymes; in fact an ideal high-energy food for the bacteria living in the rhizosphere. In normal healthy soil, as we have seen, there are 600 million bacteria per gram of soil, but in the film of micro-organisms around the plant's roots there are a staggering 1 million million per gram because of the highly nutritious food that the plant exudes.[8] The bacteria then release plant nutrients from the humus and clay particles, as well as breaking down soil minerals into water-soluble forms that the plant can feed on. So the bacteria are fed by the plants, and the plants in turn are provided with food which the bacteria have made available for them.

But that is not the end of the story. As the soil warms up in the spring and the days lengthen, the plant becomes more active, requiring more food. This increase in activity produces more root excretions from the plants, along with the increase in soil temperature making the bacteria become more active—which provides more food for the plant just when it needs it. And there is more. The dense film of bacteria around the roots, in combination with kilometres of fungal threads (hyphae), forms a shield around the roots of the plant, protecting them against disease-causing organisms that try to attack them.[9] The essence of ecology is the story of symbiotic relationships. No individual species can be seen in isolation—they have all evolved together; they all make up a dynamic and evolving community.

The most intriguing of all these bacteria is the specialist rhizobia family, which have evolved over millions of years in symbiotic association with leguminous plants—members of the pea and bean family. The rhizobia live in adapted nodules on the roots of the plants and have the ability to fix atmospheric nitrogen, not only for their own benefit, but for the plant as well, in exchange for carbon compounds which they use for energy.[10] For millions of years these bacteria have not only been vital to the plants they associate with, but have enabled their host plants to live in close proximity with other plants without having to compete with them for precious nitrogen. Not only that, but surrounding plants benefit when the roots die back in the autumn or when the tops of the plants are grazed off.

Throughout the plant kingdom we see this relationship between leguminous and non-leguminous plants. These bacteria have helped create one of the most successful of all plant families. From huge jungle trees in the Amazon to acacia trees in the African savannah and clovers growing with grasses in an English pasture, a whole family of plants trades nitrogen with its neighbours, and all because of bacteria that decided to join forces with them many aeons ago. This kind of co-operative behaviour is seen again and again in the marketplace that is the complex soil ecosystem.

As in the macroscopic world, in this microscopic environment there are controls on the overpopulation of organisms. The bacteria have predators which are essential to maintain a balanced population, in this case other bacteria, bacteriophages and protozoa.[11] Again and again we see the checks and balances that are an inevitable part of a complex and healthy ecosystem.

Mycorrhizae

For many years those most interesting and valuable of all soil organisms, mycorrhizae fungi and their fungal threads (hyphae), were largely overlooked, but the more we discover about them, the more we realize how valuable to the soil ecosystem and plant nutrition they are. Their importance in plant nutrition and the plant's ability to resist disease was originally underestimated. Not only are they important to the health of the soil and the plants growing in it, but they are sensitive indicators of whether a soil is reasonably healthy in the first place. Along with earthworms, the presence of mycorrhizae fungi in the soil tells the farmer whether he or she is doing the right thing, and helping to maintain a healthy, living and vibrant soil. For both earthworms and mycorrhizae to thrive, the soil has to be rich in organic matter. The mycorrhizae are exceedingly sensitive to fertilizers, pesticides and of course fungicides, and are the first organisms to be inhibited with conventional farming techniques.[12]

The first role that fungal mycorrhizae perform is to break down straw, tough plant cellulose and the even tougher lignin (woody material). The hyphae penetrate the straw and woody material, softening them up, which allows the bacteria to do their work in the next stage of breaking down the material into stable humus, at the same time as releasing valuable nutrients for the plants growing in the soil.

Secondly there are specialist mycorrhizae which penetrate plant roots, developing similar symbiotic relationships with the host plants that rhizobia bacteria make with leguminous plants. The mycorrhizae feed on exudates from the plant roots, as the bacteria do, and in return supply nutrients obtained by breaking down cellulose and lignin remains at the other end of the hyphae. These threads extend up to 4 cm (1.5 in) into the soil, increasing the plant roots' surface area by up to ten times. The mycorrhizae and the plant communicate in a biochemical language that enables the plant to order what and how much nutrient it needs at any one moment. In exchange for this service, the mycorrhiza is supplied with more nutritious root exudates. The plant in its turn then feeds on the nutrients that the mycorrhizae threads, embedded in its root cells, have piped back.[13] Sir Albert Howard said:

> "The mycorrhizal association therefore is the living bridge by which a fertile soil (one rich in humus) and the crop are directly connected and by which food materials ready for immediate use can be transferred from soil to plant." [14]

The brassica family and some beets, on the other hand, do not make mycorrhizae associations. As we understand more and more about plants and their associations with mycorrhizae fungi, however, it is beginning to be realized that a high proportion of species benefit from such associations. For some plant species, such as orchids and Scots pine trees, these root associations are essential for their survival. Even leguminous plants, which already benefit from the rhizobia bacteria in their root nodules, also have mycorrhizae associations—so much so that when, in one experiment, lucerne was inoculated with both mycorrhizae and rhizobia bacteria, the yields doubled, as compared with the plants which were only inoculated with their usual host, rhizobia.[15] Newman Turner observed that a healthy crop of field beans could only be grown with good quality compost. He also used to follow no-ploughing techniques. The combination of these practices would have encouraged a good growth of mycorrhizae.

The Living Soil

There is increasing evidence that plants living with a healthy mycorrhizal population are richer in minerals such as phosphate, and are healthier and more able to resist disease. In comparison to plants without mycorrhizae root associations, they have been shown to have three to ten times more nutrition in their leaves, fruit and grain.[16] The mycorrhizae threads are much more efficient at absorbing water and plant nutrients than roots on their own. These are the sort of phenomena that the Swiss scientists observed on the organic plots we discussed in the last chapter.

Once again we see the trading and co-operation that is an essential part of soil life. The bacteria that live in the root zones of plants exchange benefits with the plants, the fungal mycorrhizae do the same, and in legumes the rhizobia take up residence in the plant roots for the same mutual benefits. No species evolved alone; they all developed together as a flexible community over millions of years. Not only did the characteristics of individual species evolve, but the relationships they made with other species also did so. Without both the bacterial and mycorrhizal root associations, plants cannot have a healthy diet and their immunity to disease is severely compromised.[17]

Algae

Most of the algae found in soils live on or near the surface. Like the higher plants, blue-green algae contain chlorophyll and are therefore able to photosynthesize. They can also fix atmospheric nitrogen, thereby continuously adding to the total nutrient status of the soil.[18] Moreover, they have a binding effect on any exposed soil surface, protecting it to some extent from erosion and also, presumably, from nutrient loss. I have seen them in our fields on the bare soil after a warm, damp summer like a threaded mat holding the surface soil together.

Soil Animals

Soil animals, including centipedes, millipedes, spiders, mites, springtails, larvae of various sorts, wireworms, ants, snails, slugs and earthworms, voles, mice and moles, all have their part to play. Some eat plants and their remains; others, like centipedes, spiders and beetles, eat the plant eaters, keeping them in check, and most of the larger ones help to aerate the soil with their burrowing. Here, however, I am going to concentrate on the earthworms, as they are in my opinion the most important and interesting of all the larger creatures that inhabit the soil.

Dr Stewart and the Earthworm

In the late 1970s I attended a lecture by Dr V. I. Stewart of Aberystwyth Agricultural College, University College of Wales. The late Sam Mayall from Shropshire was one of the first British organic farmers, and every year a lecture is held in his name. It was an inspirational tour de force. Dr Stewart's life's work had been the study of soil science, and in particular soil structure and the role of earthworms in helping to create and maintain it. The earthworm is one of the most important factors in creating and maintaining both soil crumb structure and soil fertility. It is basically made of rings of muscles with a mouth at one end and an anus at the other. In between it has a digestive system with a few interesting additions, such as a gizzard containing grit for grinding up food like a bird, and a special gland that secretes lime. As it burrows, it ingests soil, organic matter and grit, and with the help of bacteria and enzymes in its digestive system obtains its food. In the process it mixes the soil constituents, at the same time turning nutrients that were unavailable to plants into available water-soluble minerals, and turning plant remains into more valuable humus.[19]

And they do shift some soil! Estimates suggest that they swallow, process and excrete 40 tonnes of dry earth per hectare (16 tons per acre) every year, which is not surprising because the number of worms in a healthy British pasture are in the order of 1,250,000 per hectare (500,000 per acre), weighing 650–1100 kg per hectare (580–980 lb per acre), although 6 million were found in a pasture at Wye in Kent, weighing nearly 1.7 tonnes.[20] *Lumbricus terrestris* (the common earthworm), the largest British worm, pulls leaves and other organic matter down into its burrow to digest, helping to incorporate organic matter into the soil. Others, like *Allolobophora caliginosa*, on the other hand, tend to feed on organic matter below ground, depositing their casts on the surface, thus building up a fine topsoil and carrying nutrients that have washed out of the topsoil back to the surface.[21] The fine topsoil builds up over many years, and the larger stones find their way to the bottom, just above the subsoil, thus helping to provide a drainage layer at the base of the topsoil. Others deposit their casts within the soil, helping to create an essential stable crumb structure. Every morsel of topsoil has passed through a worm's digestive system numerous times. Brandling worms or tiger worms (*Eisenia foetida*) are at the start of the humus food chain, and they must also not be overlooked. They are found in cow pats, compost and muck heaps and in the leaf litter of woodlands—in fact wherever there are high levels of reasonably fresh organic waste. These worms are the first in line when it comes to the process of breaking down organic waste into plant food.

Although earthworms will not survive in very acid soils, they have the ability to maintain a neutral soil by excreting calcium surplus to their requirements from a special gland in their gut. In other words, the earthworm, whilst helping to create and maintain the environment in which it likes to live and prosper, also benefits a wide range of other soil organisms and plants. Some of the calcium they ingest is insoluble, but during digestion it is converted into a soluble form, thereby reducing its acidity, and helping to maintain the essential crumb structure of a healthy soil. This is because calcium ions have a positive electrical charge which attracts the negatively charged particles of clay which coagulate around the calcium particle, forming a crumb. The gut of the earthworm is the ideal environment for soil bacteria to proliferate, and therefore the worm's casts are much richer in bacteria than the surrounding soil.[22]

To sum up, worm casts are much richer in available plant nutrients, have some easily available calcium, are richer in more valuable humus and have a much higher bacterial content than the original ingested soil, making them rich in the ideal plant foods. Moreover, the worm adds to this mix a sticky mucoprotein which glues the soil particles together, forming a crumb which is able to resist the eroding action of rain.

To reinforce the walls of their burrows, earthworms line them with their own cast material, compressing it by expanding their bodies—and plant roots love the results! In their search through the soil, they find an aerated tube with easy access and plenty of available food. And when it rains, the water runs conveniently down the tunnel to water them—what more could a plant want? But it is even better than that.

What Dr Stewart did in his research was to calculate the ideal soil crumb size. He considered that there are two main factors involved. The first is the air to water ratio in the soil, which in turn is determined by the surface area of the crumbs in relation to the air spaces between them, and the nature of water's surface tension. After rain, when the excess water has drained out of the soil, some water remains as a film on the surface of the crumbs. If the crumbs are too large, there is too much air and not enough water. If they are too small there is too much water—indeed, there may be no room for air at all. The other factor is that if the soil crumb was too big, oxygen could not penetrate to its centre, which would become stagnant and anaerobic, which is not ideal for plant roots. Using these criteria, Dr

Stewart calculated the perfect crumb size, and found that it corresponded exactly to the size of the crumbs created by worms!

Banked and Invested Nutrients

There is another side to this story. We have seen how the soil is home to a huge and complex collection of micro-organisms, a large number of which help to convert organic matter and non-available minerals into water-soluble chemicals that plants can feed on. In other words, these minerals are 'banked' or 'stored', ready to be released when necessary. If they were all in water-soluble form, they would have been washed out millions of years ago, and plants and the soil would have become a barren desert.

However, plants do need their food in water-soluble form, as Liebig, the father of agricultural chemistry, discovered. By burning plants and analysing the remains, he concluded that the minerals that were left were those that plants needed for good growth. The three main elements he identified were nitrogen, phosphorus and potash. As I have said, the oversimplification of his discoveries has led to many of the problems of modern agricultural practice, but they are very important in the context of a living soil. It is how these nutrients become water-soluble that is important to the understanding of plant nutrition which underpins organic agricultural practice: "Feed the soil, not the plant." Feed the micro-organisms: the mycorrhizae and other soil life provide the right conditions, and the nutrition of the plant will be taken care of automatically.

In healthy soils there is a huge store of nitrogen, phosphorus and potash, as well as other nutrients, but mostly in an unavailable form. They are released by the micro-organisms when the conditions are right, but the manner and the timing of their release is fascinating.

There is no nitrogen in the underlying rock component of the soil, so what is its source? A small but significant supply comes from the action of lightning, which combines nitrogen and oxygen to form nitric oxide in the atmosphere. This in turn is dissolved in rain and descends to earth in the form of dilute nitric acid, which is converted to nitrates by soil organisms. Some is fixed from the air by a variety of soil micro-organisms, including rhizobia, but the bulk comes from animal and plant waste which arrived in the soil in the near or distant past, and has continuously been recycled ever since.[24]

However, a sizeable amount of the nitrogen in organic waste is unavailable. Humus, therefore, is the first example of plant foods being placed in the 'bank'. The second is inorganic nitrogen, which is used as food by soil microbes to build their bodies, thus locking it up temporarily. A third is free ammonium ions in the soil produced by micro-organisms breaking down humus. These have a positive charge and are attracted to clay minerals which have negative charges, so they are locked up by the clay particles.[25] There is therefore nitrogen at different levels of availability in the soil at any one time: ready for a plant to use immediately; unavailable but easily released; and in a form more difficult to release.

Potassium, on the other hand, does not arrive from outside; it is there to a greater or lesser degree in the soil, clay and underlying rock. Of course, it is returned time and time again in the form of decaying plant and animal material, but it came from the rocks in the first place. In a natural system it cannot be added to, so the mechanisms for stopping its loss and the control of its release have to be much tighter than those for nitrogen. In most soils there is an almost inexhaustible supply: the only problem may be its availability. It is very quickly locked up by clay particles, but is very slow to be released. So what are the mechanisms for making potassium available to plants? It is a four-step process, with a form of less easily available potassium changing into a more accessible one. The four forms of potassium are:

1. Potassium in soluble form, immediately ready for plant use.

2. Potassium in an 'exchangeable' form, which is released as and when the potassium in solution becomes depleted. This is the beauty of the system: the release of potassium from its 'exchangeable' form is triggered by the plants' roots as the level of available potassium is reduced. In other words, it is released when the plants need it. This means there is the minimum amount of potassium at any one time that could be washed away by the soil water. There are different levels of availability, as with soil nitrogen, but the mechanism of potassium release is a much more tightly controlled system.

3. Potassium in a 'fixed' form, which can only be released when the levels of 'exchangeable' potassium are critically low.

4. Potassium in the 'mineral' form, which is only transformed into the 'fixed' form very slowly by weathering and soil bacterial activity.

This is such a tightly controlled mechanism that often when soil tests are done on organic soils or in natural conditions little or no water-soluble potassium is detected, although the plants show no signs of potassium deficiency, because they are receiving potassium in solution only as they need it.[26]

The plants trigger the release of just enough potassium ions from the 'exchangeable' form for their needs, by encouraging bacterial activity in their root environment as described earlier. This in turn depletes the 'exchangeable' stocks, causing the release of 'fixed' potassium, which is eventually restocked from the 'mineral' form. Thus after millions of years we still have an almost inexhaustible supply, trebly protected against loss.[27]

However, there are areas, such as parts of Australia and the Amazon basin, where potash supplies have run low due to geological stability over billions of years, which has led to supplies being slowly leached away by the action of rain over the millennia. Over most of the Earth's surface, however, where volcanic activity has been more recent, or where old sedimentary rocks have been forced to the surface by geological folding, nutrients such as potassium are renewed on a comparatively regular basis.

Phosphate

There is also a bank of unavailable and low-soluble mineral and organic forms of phosphate in the soil. It is fascinating that, as with potash, there often appears to be a deficiency of phosphate in the soil despite the plants showing no signs of deficiency, because it is being released only when it is needed. It has been shown that the soluble phosphate content in the soil will remain approximately constant whilst a crop is growing despite the fact that it is taking up considerable amounts from the soil. In other words, as quickly as the roots absorb phosphate from the soil solution, more is made available from the store.[28]

As with potassium, certain species of micro-organisms are involved in this process, converting low-soluble phosphate into its soluble form, triggered by the plants' roots as and when they need it. But probably more important than the micro-organisms is the part that mycorrhizal fungi play. They are very efficient at converting phosphate into a soluble form and making it available to the plants by penetrating their roots and trading the phosphate and other minerals in exchange for carbohydrates. Phosphate also plays an important role in helping the plant resist disease.[29]

Nature's Agriculture

Although I have had to simplify the complex world of the soil's ecosystem, I hope I have given you some idea of this fascinating living environment, with its mixture of minerals, ground rock, humus and micro-organisms

The Living Soil

which together help to create the ideal conditions for plant growth.

I started this chapter with a quote from Sir Albert Howard about how we should study Nature's soil management in order to farm and garden better. He continues with an eloquent description of the natural processes at work in a woodland or forest:

> "What are the main principles underlying Nature's agriculture? These can most easily be seen in operation in our woods and forests.
>
> Mixed farming is the rule: plants are always found with animals: many species of plants and animals all live together. In the forest every form of animal life, from mammals to the simplest invertebrates, occurs. The vegetable kingdom exhibits a similar range: there is never any attempt at monoculture: mixed crops and mixed farming are the rule.
>
> The soil is always protected from the direct action of the sun, rain, and wind. In this care of the soil, strict economy is the watchword: nothing is lost."[30]

How then can this understanding of natural processes be translated into farming practice? If we follow the basic principles below, we cannot go far wrong.

1. Create as much biodiversity as possible, both in and above the soil. This is done by a regular supply of organic matter, and keeping a good pH. This will lead very quickly to a thriving soil life, which will provide all the crops we need.

a) Above the soil, create and encourage as much diversity as possible, in both crops and livestock.

b) Sow complex mixtures of grasses, clovers and herbs in the pastures.[31]

c) Have regular rotations of crops and animals.

d) Intercrop two or more compatible crops together, as is done in Latin American countries where 80 per cent of cereals are regularly intercropped with up to twenty species grown in close proximity, and 60 per cent of maize is intercropped with beans.[32] In Cuba, cassava-beans-maize, cassava-tomatoes-maize and sweet potatoes-maize mixtures have proved 1.5–2.8 times more productive than the sum of individual monocultures.

e) Use weeds productively, in a controlled way, to act as food for natural predators, to help protect the soil and act as green manure.

f) Allow nature reserve areas and healthy field margins.

g) Create or preserve ponds.

h) Have areas of mixed woodland.

All these practices will produce a healthy and robust ecosystem.

2. Protect the soil. The soil should be protected from the sun, rain and wind as much as possible with top dressings of compost, mulches, ground cover, green manures and the judicious use of weed cover, as well as the crops themselves.

3. Encourage a good crumb structure. This is done by regular supplies of organic matter, which encourage all the micro-organisms, worms, etc., that help to create a healthy crumb structure. A policy of little or no tillage, digging or ploughing, so as not to disrupt the soil profile, the soil network of drainage and the aeration channels made by earthworms and other burrowing animals, will also ensure that the crumb structure reaches its optimum. Occasional subsoiling, or aerating with a fork in a garden, will help on heavy soils, but once a good open soil profile is achieved, it will be less necessary.

4. Return all organic waste. As in Nature, this is best done by adding it to the surface, or at most, lightly mixing it in. Let the army of worms and other creatures incorporate it in the soil. I used minimum cultivations on my farm as time went by, and the soil structure improved and the worm population increased. Compost and green manures were incorporated into the top few inches of soil, rather than ploughed under whenever possible, and if ploughing was done it was as shallowly as possible. We have not dug or inverted the soil in our vegetable garden raised beds for at least twenty years and the topsoil gets more and more friable and sweet-smelling every year. If we do use a fork, it is to loosen the soil occasionally by wriggling it as one aerates a lawn. And when we inspect the soil beneath, it is like a sponge, full of spaces and worm holes, and with a huge population of worms which thrive in such undisturbed conditions.

5. Use deep-rooting herbs and crops whenever possible. Deep-rooted herbs in grass mixtures, such as chicory and salad burnet, green manures such as field beans and lucerne, and vegetables can tap the subsoil, bringing leached minerals back to the surface as well as mining for as yet unavailable ones. Thus we are recycling much of the minerals lost to the surface layers of the soil and to crops with shallower roots, in a similar way that trees recycle minerals in the forest.

Healthy Soil, Healthy Plants, Healthy Animals

If a farmer follows the above principles, the whole ecosystem becomes healthier, and the plants and animals are more able to resist pests and disease. The converse is also true. There are increasing examples of the combination of monocropping and pesticides intensifying predator attacks on crops. The most remarkable example was discovered by Peter Kenmore and his colleagues in the 1980s in South-East Asia. They found that the higher the doses of pesticides in rice fields, the more numerous were the pest attacks, because the pesticides were killing off the pests' natural predators. This led to the Indonesian government banning fifty-seven types of pesticide in 1986 and setting up field schools to educate farmers about the benefits of biodiversity and integrated pest management (IPM) schemes.[33] These practices have spread to other countries, like Vietnam. What has been found is that most of the time rice can be grown without the application of pesticides and without a reduction in yields, as long as biodiversity and IPM are maintained.

Pesticides also result in increased natural resistance in the pest or disease. Some 480 species of insects, mites and ticks are recorded as having become resistant to one or more compounds, 113 weed species have become resistant, as have 150 fungi and bacteria.[34]

Whenever my wife gives a talk about organic gardening, there are always the inevitable questions about how we deal with pests and diseases. She often finds it difficult to convince her audience that, although we do have them, once an organic system has been established for a few years, the incidence of disease diminishes, and when it does occur it tends to be less overwhelming. Needless to say, the sceptical audience will often ask about organic alternatives or acceptable organic insecticides, because they are still thinking in terms of curing the symptoms, instead of creating healthy conditions and preventing pests and disease in the first place.

I can honestly say that we have used hardly any insecticides over the years. At the start we occasionally used Derris or Pyrethrum, and more recently aluminium sulphate to control slugs. Currently we use the biological control *Bacillus thuringiensis* against cabbage white butterfly and cabbage moth caterpillars. It is sprayed on the leaves and is ingested by the caterpillars, which gives them a terminal case of dysentery; but it is safe for other butterfly larvae. On field crops we never used anything. If potato blight struck, we caught it early and flayed off the tops, leaving extra time before harvesting to allow the spores to die.

However, for those who are still sceptical, let us briefly look at the kind of emergency methods that are allowed on organic establishments, before we go on to examine how to encourage healthy farms, soil, plants and animals which can resist disease.

Biological and Other Controls

Controls of diseases and pests include naturally occurring plant extracts, minerals and biological controls. The value of liquid extracts is that they are natural substances which, although not harmless to the pest or disease being treated, are comparatively benign in the general sense and break down leaving no dangerous compounds to accumulate in the environment to the detriment of plant, animal and human life.

Liquid extracts of compost have been used for fungal disease control. This is not new; there is evidence that the ancient Egyptians used compost and manure 'teas' on crops 4,000 years ago. There have been several research studies into compost extract use and its effectiveness in Germany, Britain and the USA. These show that compost teas have natural fungicidal properties. The active ingredients are live bacteria in the genera *Bacillus* and *Serratia*, as well as fungi in the genera *Penicillium* and *Trichoderma* and others. Compost teas have been shown to act in three main ways:

1. Inhibiting spore germination;

2. Antagonism and competition with pathogens;

3. Induced or acquired systematic resistance against plant pathogens.

Compost teas are made by steeping compost in water, to which is added extra food, such as molasses. The liquid is kept aerated to encourage virulent bacterial growth.[35] Needless to say, when it is sterilized its effectiveness is greatly reduced.

Plant extracts can be used to reduce incidences of damping off in seedlings by treating the seeds for the control of fungal diseases and insects. Plant extracts have long been used by organic horticulturists for their fungicidal effects. These include valerian, chamomile, horsetail, horseradish, garlic, onion, wormwood, dock, stinging nettle, rhubarb, neem and quassia, among others.[36] Minerals such as waterglass are also used to strengthen plants against insect attack.[37]

Biological controls such as *Bacillus thuringiensis* (already mentioned), the predatory mite *Phytoseiulus persimilis* for the control of red spider mite, and the chalcid wasp to control whitefly in greenhouse conditions, can all be used when necessary.[38] In Africa the cassava mealybug has been controlled by releasing predatory wasps by air.[39] Biological controls are used in conventional systems, but their use in organic farming is seen as a last resort. There are already worries that insects will become resistant to *Bacillus thuringiensis* owing to its overuse.

Mechanical controls such as netting, vertical barriers against carrot and cabbage flies, traps, and sound for bird scaring, have been found to be very effective. Before we used barriers against carrot fly, we always had a level of infestation. Because the carrot fly hunts close to the ground, seeking a host carrot, a 70cm-high barrier around the plot foils the flies. Alternatively, one can grow them under fleece, which does the same job.[40]

For livestock, antibiotics are not used routinely on organic farms, but can be used in emergencies. Herbal remedies are commonly used, with their long period of efficacy. Some stockmen use homeopathic medicines and other complementary approaches. Vaccines are used, but again, as with antibiotics, it is better to remedy the conditions which give rise to the problems in the first place and use them sparingly.[41]

Heyam Dukam Anagatum ('Avert the danger before it arrives')

This Sanskrit aphorism says it all. A large part of the answer to pests and disease is not how to cure, but how to prevent. That is not to say that they do not get out of hand every now and again in any organic system—and when they do, emergency measures have to be taken—but starting from the premise: 'Wait until nature strikes, then zap it with some miracle cure' is to completely miss the point. Contrary to popular opinion, Nature is not out to get us. To quote Newman Turner: "I often think it is man's desire to destroy that creates within him the fear which gives rise to the belief that nature has destructive intentions against man."[42]

We have already seen many instances of natural mechanisms that keep pests and diseases from getting out of hand. One example is the increasing evidence that plants are more able to resist disease when living with a healthy mycorrhizae population. None of these examples adds up to much on its own, but when taken together with all the other elements of any established mixed organic system, they make a big difference.

We are always hearing from ecologists about the importance of diverse ecosystems. Indeed the loss of diversity is seen as a sign of the degradation of that system. The stability of a system is in direct proportion to the number of interacting species within it. But it is far more fascinating than that. It is not natural for a system to be static; the stability we are talking about is a flexible, dynamic stability that has the ability to adapt and change according to changing circumstances, from week to week, from season to season, from year to year.

How does this come about? Let us take a look at the soil ecosystem. In our culture, we tend to see things in terms of 'baddies' and 'goodies'. We see the 'bad' pests and diseases on the one hand, and the 'good' predators of those pests along with 'helpful' organisms like earthworms on the other. This is another example of a lack of holistic vision and understanding. Here is Newman Turner again:

> "If it is not blatantly obvious that a plant or an animal or any other phenomenon of nature has a value to our commercial activities, then we attempt its destruction without further thought. If anything appears in the least way to obstruct, or indeed fail to serve, our artificial activities, our main desire is to be rid of it—to remove it from the face of the earth. It is this flaw in human intelligence which has allowed us to destroy vast areas of fertile land and, in a smaller way on our own British farms, to bring upon ourselves untold pests and diseases which would have remained under the control of nature had we not thoughtlessly destroyed that part of nature whose purpose it was to control the pest or disease." [43]

In a natural system, pests and diseases have a specific role. For one thing they attack weak points in the system, making space for healthier organisms or better-adapted species. Some of them are involved in breaking down weak or dead organic material under normal circumstances, only becoming pests when soil and plants become unhealthy. Under healthy conditions they are kept under control by predators.[44]

This tendency to see only good or bad is a European cultural failing. In the East it was traditionally understood that both destructive and constructive forces are just two sides of the same coin. Destructive forces are only bad if the system becomes unbalanced and they begin to dominate, but equally, creative forces themselves can be life-threatening if they are out of control. When the system is in balance, dissolution and creation are but part of the same evolutionary cycle, as in the rotting down of dead plant and animal material to provide humus and food for a new generation. Seeing complex ecosystems from this perspective we have a vision of a soil society and general environment with checks and balances. Every species has its role and value in the totality. No disease or pest is allowed to let rip, no species in the system is allowed to dominate. As Sir Albert Howard said:

> "The crops and livestock look after themselves. Nature has never found it necessary to design the equivalent of the spraying machine and the poison spray for the control of insect and fungous pests. There is nothing in the nature of vaccines and serums for the protection of livestock. It is true that all kinds of diseases are to be found here and there among the plants and animals of the forest, but these never assume large proportions." [45]

In short, here is a community that exists in dynamic equilibrium, competing and co-operating with its checks and balances. Some species become more numerous when changing circumstances require it, others come into play when the circumstances change, which causes the whole ecosystem to settle down to a new equilibrium, more suitable for the new circumstances. This is the great strength of a complex system: its strength lies in its flexibility, which comes from its diversity.

Equivalents

In a healthy farming system, the purpose therefore is first to encourage a healthy and vibrant ecosystem which to a large extent can look after itself; and where agricultural practices limit the kind of diversity seen in a wild system, equivalents are provided. An example mentioned in the last chapter is crop rotations. A crop is not grown more than one season in the same place, but alternates with complementary crops. This means following each crop with one that in some way benefits from its predecessor, which helps to stop the build-up of pests and diseases specific to a particular crop. If we grow one crop on the same ground season after season, those nutrients that a particular crop extracts from the soil become depleted and the plants become weak and more open to infection.[46]

A good example of complementary crops are field beans followed by potatoes or wheat, or in gardening, peas and beans followed by brassicas. In both cases the preceding crop of legumes which have nitrogen-fixing nodules on their roots help to feed the following nitrogen-hungry crops. Also, in the case of the field beans and potatoes, the field beans have deep roots (up to 1.2 metres/4 ft) which have a very beneficial effect on the soil by opening it up and improving its structure, which benefits the potatoes as they like an open, crumbly soil. These practices help to keep the plants healthy and improve their resistance to disease.

Another version of this idea, which is closer to natural systems, is to grow more than one crop together, either in alternate rows or strips or mixed together. There are many examples of this around the world, both currently and from the past. Much research has been done into the benefits of mixed cropping for the control of pests and diseases. In his book *Organic Farming*, Nicolas Lampkin cites eighteen examples of intercropping which have been shown to have a beneficial effect in controlling the insect pests of one or both of the companion crops.[47]

Yet another version of diversity is to have a mixture of different varieties of cereal growing in the same field, or a mixture of different species, such as winter wheat and field beans or spring oats and peas. This is a rediscovery of an old practice, but none the worse for that. Professor Martin Wolfe and scientists at the National Institute for Agricultural Botany have experimented with mixed varieties of the same cereal as well as legume and cereal mixes, both of which showed a reduction in diseases such as powdery mildew. There was less cross-infection simply because the individual plants were further apart, separated by the companion crop, or in the case of mixed varieties of cereals, because the different varieties were susceptible to different strains and types of disease.[48]

This has also been borne out by recent research done in China. In 1999, a team of Chinese scientists tested one of the key principles of modern rice growing. They planted hundreds of hectares of a single high-tech variety of rice at the same time as hundreds of hectares using a much older technique: planting several older breeds of rice together in the same field. To the scientists' amazement, the mixed plantings yielded 18 per cent more rice than the high-tech crop. Not only that, but the devastating fungus rice blast, which usually attacks monocropped rice, and which requires repeated applications of pesticides to control it, was reduced by 94 per cent. As a result they found they could stop applying pesticide almost entirely for the mixed crop.[49]

The same maxim applies to the rotation of animals. Every stockman knows that apart from different kinds of stock utilizing different parts of the grass mixtures, pests such as parasitic worms tend to be reduced if the types of stock are alternated, breaking the natural cycle of pests that are specific to that type of animal.

Susceptibility to Plant Disease

Over the years there has been much research into the increased susceptibility of plants to infestation by both insect pests and fungal diseases in those crops grown with mineral fertilizers, as compared to crops grown with composts or manures. I will oversimplify the explanation slightly to make it more understandable. In crops grown with fertilizers, there is an increase in the water, sugar and protein content of the plants' cells, and a decrease in the cell wall thickness. In other words, the plants become obese and frail. As a result it is much easier for sucking insects such as aphids to penetrate the cell walls, to which they are attracted in the first place because of the increased nutrients. They also multiply more quickly because of their increased food supply and because of the thinness of the cell walls or the damage done by the insects. Moreover, because of the increased sugar content, fungus diseases proliferate. Needless to say, the opposite is true for organically grown crops.[50] As we have seen, mycorrhizal fungi threads are much more prevalent in healthy organic soils, and their association with a plant's roots makes the plant healthier and more able to resist disease.

Recent research at the University of California also shows that pesticides and herbicides thwart the production of phenolics—chemicals that act as a natural defence for plants.

Weeds

One area that I have not yet mentioned is the control of weeds, especially in cereal crops, which are difficult to hoe once the crop is growing. Many people assume that one cannot grow organic crops without an infestation of weeds. They are so used to using sprays and believing the chemical manufacturers' propaganda, that they cannot see how weeds can be controlled. However, their grandfathers would have been able to tell them about simple, tried and tested methods to keep their crops clean. Barry Wookey, an organic cereal and beef farmer from Wiltshire, described in his book *Rushall: The Story of an Organic Farm* how he prepared his fields for autumn sowing by using the 'false seed bed' technique, which involved preparing the ground ten to fourteen days before sowing took place. The weeds then grew, but when they were still in delicate early growth the subsequent mechanical actions of drilling and harrowing eliminated the majority of them, as well as left-over cereal seeds from the last crop. As he said, the visitors to his farm were always surprised to find how weed-free his cereal crops were.

Newman Turner had the same results, largely because he used a no-plough technique which involved disc harrowing the top few inches and killing the young weeds in the same way. Farmers around the world who use no-tillage techniques are also controlling weeds, because the weeds are largely smothered by the green manure and crop residues left on the surface of the soil. On a garden scale, this is the kind of technique my wife uses to suppress weeds—with mulches. She uses mainly grass mowings in summer and straw in the winter, both to great effect.

Rotations also help to control weeds, as we discovered ourselves with our seven-course rotation. This comprised four years of herbal-grass-clover ley, followed by three years of arable—wheat and/or barley, followed by field beans, and finally root crops including potatoes, fodder-beet, swedes and kale. Annual weeds would germinate and die out during the four-year ley, and the perennial weeds such as thistle, dock and couch would be easier to deal with, especially when growing and harvesting the root crops. The constant switch from temporary grass leys to arable crops helped to keep on top of the weeds.

Be a Student of Nature

So the key to farming productively, healthily, and sustainably is to study Nature and how she does it, then follow her example. There is an idea prevalent in some quarters that everything we do is unnatural, as though we are somehow not part of Nature. However, if we learn from Nature and follow her laws in our agricultural practice, the results will be life-supporting and genuinely productive, rather than destructive. As always, the choice is ours.

Chapter 4

Nutrition and Food Safety

"These pesticides are now big business, with global sales exceeding US$31 billion in 1998. Each year, farmers apply 5 billion kilogrammes of pesticides' active ingredients to their farms."—Jules Pretty[1]

Food Quality

The increasing consumer interest in organic food and the huge growth in production and sales in the West, together with the recent growth of organic and sustainable agriculture in Third World countries, are extremely hopeful signs. The thought, therefore, of writing about the worst aspects of conventional farming was daunting. My inclination is to concentrate on the positive, to tell the ancient history of sustainable farming and the story of the modern organic movement and the life-affirming ideas and theories behind it, but I would not be doing justice to the subject if I did not tackle the things that worry most people about present farming practices: the pollution of both our food and the environment, BSE, antibiotic resistance, and genetically modified food.

The Food Standards Agency
The Food Standards Agency (FSA) was set up on 3 April 2000 as a separate organization from the Ministry of Agriculture because of the clash between farmers' interests and consumers' interests following the BSE crisis. The FSA was until recently headed by Sir John Krebs, a zoologist by training. He has been a research scientist all his life and worked for the then Ministry of Agriculture before being appointed to the Food Standards Agency, and is a passionate proponent of the modernist scientific approach to agriculture. The following Christmas it was reported that more people had bought

organic food for the holiday season, in Sir John's words, "despite the fact that there was no evidence that it was healthier." He continued: "They're not getting value for money, in my opinion and in the opinion of the Food Standards Agency, if they think they're buying food with extra nutritional quality or extra safety. We don't have the evidence to support these claims."

His scepticism about organic farming was not surprising, but his claim that there was no evidence was simply incorrect. There have been numerous studies comparing organically and conventionally grown food over the years, which show that organic food is in many instances measurably more nutritious, and definitely less contaminated with pesticides. There is also growing evidence of the hazards of ingesting even very small amounts of pesticides in our food, particularly when more than one is present. Indeed in 2001, the Soil Association collected together over 400 scientific papers on the quality of conventional and organically grown food. This they have published in a report entitled 'Organic Farming, Food Quality and Human Health'.

Originally I felt prepared to give Sir John Krebs the benefit of the doubt, assuming that he did not know about the research, but I subsequently discovered that he has stated publicly that he was trying to 'undermine' claims that organic farming was more environmentally friendly than non-organic agriculture. The Food Standards Agency claims that it wants to reduce pesticide residues on our food, but at the same time refuses to promote organically produced food, which is the simplest way to achieve this.

These comments by Sir John Krebs about the equivalence of organic and non-organic food were made after the BBC had commissioned a test on just three carrots! Two had been grown organically, one in Britain and one abroad, and one conventionally. These were tested by the Eclipse Scientific Group laboratory in Cambridgeshire, England, and all three were found to be free of forty different pesticides residues known to be associated with carrot production.[2] In scientific circles this would normally be considered such a ridiculously small study as to be of no significance at all, and to make such a claim on a study of only three carrots, without reference to all the other research on this subject in this and many other countries over the years, is both unscientific and misleading. The Central Science Laboratory's most recent report on pesticide use shows that on average, carrots grown conventionally are treated four times during their growth with different insecticides, three times with herbicides, and three times with fungicides. And as

Sarah Hardy, the Soil Association's Producer Technical Manager has said, "The high use of chemicals and the fact that carrots act like sponges for chemicals, is the reason why I say to people, if you eat nothing else organically, eat organic carrots."[3]

Sir John Krebs also said, "I think the organic industry relies on image, and that image is one that many consumers clearly want to sign up to." I presume this means that he thought the increasing number of consumers who buy organic produce are just following a fashion that in time will die out.

Many people were saddened by these comments, because of the excellent work that the Food Standards Agency does in trying to protect consumers by looking into problems of food quality. Most people assumed that it was going to be truly independent of the Department of the Environment, Food and Rural Affairs (as it is now called) and the agriculture industry's vested interests. This they undoubtedly are when it comes to reporting on possible health problems, misleading labelling etc., but when 31,000 tonnes of pesticide-containing liquids are used every year in the UK, and around 400 tonnes of antibiotics were used in agriculture in 2001 alone, the public have a right to be sceptical about suggestions it does not get into our food, or have any repercussions for our health.[4] Many members of the public are increasingly irritated by being treated as unintelligent and are becoming more and more cynical as a result. It is quite common nowadays, that when a government or other official spokesperson makes statements to the effect that our food is perfectly safe, the result has the opposite effect and the sales of organic food increases even further.

There were doubts about the Food Standards Agency being independent. Baroness Dean's 2005 report of the FSA gives the agency a reasonably clean bill of health in this respect, but it went on to say:

"There were, however, concerns that, in fiercely defending its independence, the Agency does not always pay due regard to externally-produced information and so the evidence upon which decisions are made may sometimes be too narrow."

This could possibly explain why Sir John Krebs has said what he has said about "evidence not being available".

The verdict is still out on whether the Agency is truly independent of the agricultural industry. There are also serious doubts about its lack of powers. Unless the government gives it real teeth, the public will continue to wonder whether the FSA is really there to protect consumers.

Scientific Studies into the Nutritional Content of Organic Food

Time and again we have heard experts argue that there is no difference between organically and conventionally grown food—'a carrot is after all a carrot, however it is grown'. But comparative studies have shown not only more nitrates, free amino acids, oxalates and other undesirable compounds in food grown conventionally with mineral fertilizers, but also a decrease in many of the desirable compounds. In 1975, after a twelve-year study, Schuphan found the following differences between organically and conventionally grown food:[5]

- 23 per cent more dry matter
- 18 per cent more protein
- 28 per cent more vitamin C
- 19 per cent more total sugars
- 13 per cent more methionine (an important amino acid)
- 77 per cent more iron
- 18 per cent more potassium
- 10 per cent more calcium
- 13 per cent more phosphorus

The reductions in undesirable compounds in the organic produce were as follows:

- 12 per cent less sodium
- 93 per cent less nitrate
- 42 per cent less free amino acids

Similar results were arrived at by Fischer and Richter in 1986, Lairon in 1986, Abele in 1987, Bulling in 1987 and Kerpen in 1988.

More recent studies have shown that intensive agriculture reduces and almost eliminates the mycorrhizal fungi that grow in the soil, resulting in a decrease in the plants' mineral content. The use of fertilizers has also been shown to have the effect of reducing the amount of minerals such as zinc, copper and selenium taken up by the growing crop.

At the University of California, Dr Alyson Mitchell and her colleagues have been investigating the differences in cancer-fighting antioxidants in organic and conventionally grown fruits and vegetables, including sweetcorn, strawberries and marionberries (a type of blackberry). Their research shows that significantly higher levels of antioxidants are found in the organically grown food. The research is to be extended to examine tomatoes, peppers, broccoli and other vegetables, and the research group expects similar results.[6]

Another recent study compared research done in 1916 into the iron content of apples to those of modern apples. To have received one's daily requirement of iron in 1916, a person would only have had to have eaten two apples. Today one would have to eat twenty-three! And in 1991 a report by the Ministry of Agriculture and the Royal Society of Chemistry showed that mineral levels in fruit and vegetables had reduced by between 15 and 76 per cent since 1940.[7]

In February 2004, the Aberystwyth-based Institute of Grassland and Environmental Research published a report showing that organic milk had higher levels of essential nutrients than conventional milk, including two-thirds more omega 3 essential fatty acids, which are vital for brain function and development, and help to keep our hearts healthy and combat the effects of arthritis.[8]

Flavour

One of the significant differences between organic and conventional cultivation is in the water content of fresh food such as vegetables. If Schuphan found 23 per cent more dry matter in the vegetables he studied, it follows that there was less water content in the organic food, and hence inevitably more food value and more taste. Some studies have shown that organic food has more flavour, while others showed no difference. However, with less water and more minerals, there is likely to be more flavour. In my experience, organic food does generally taste better. Having compared our Desirée potatoes with those of a conventional farmer's, the difference was quite marked.

The biggest difference, however, is in the taste of meat and eggs. Free-range eggs are infinitely better, and organic pork and lamb are unsurpassed. The difference is so pronounced as to be undeniable. We genuinely felt we had been conned all those years and denied the wonderful tastes we were now enjoying.

Denis T. Avery

The views of Sir John Krebs are moderate when compared with those of Denis T. Avery, an American with a mission to promote "high yield farming to save wild life" as he puts it. He preaches a gospel of biotechnology, pesticides, irradiation, factory farming and free trade, and says it is the 'greenies' and 'organic frenzies' who are threatening the world with famine and loss of habitat for wildlife, because farming without synthetic pesticides and fertilizers, and without biotechnology, would require too much land. He also denies that there is any link between pesticides and cancer or other illnesses.

He claims that the US Centers for Disease Control (CDC), which tracks outbreaks of food-borne illnesses, has research to show that people who eat organic foods are eight times more likely to be attacked by the deadly new strain of *E. coli* bacteria 0157:H7, because organic food is grown with animal manures, which often contain this bacteria. He continues to make these claims, despite denials from the CDC. Eventually the CDC took the unusual step of issuing a press release on 14 January, 1999, stating: "The Centers for Disease Control and Prevention has not conducted any study that compares or quantitates [sic] the specific risk for infection with *E. coli* bacteria 0157:H7 and eating either conventionally grown or organic/natural foods." In addition, Dr Robert Tauxe, one of CDC's chiefs, phoned Avery to tell him to stop claiming that the CDC was the source of the allegation. Avery responded, "That's your interpretation, and I have mine."

In fact, *E. coli* 0157:H7 contamination is less likely to occur on organic farms, because the use of raw manure is frowned upon in the best organic practices. The composting of animal waste, along with bedding straw and other high cellulose waste, tends to eliminate such contamination.[9] On the contrary, it is conventional farms, which grow crops that have access to raw manure, often from intensive poultry or pig units, that are far more likely to have these problems.

Pesticides in Food: The Research

"Some studies have been published which claim that there is in fact no difference between organic and conventional produce as far as pesticide residues are concerned. On closer examination, however, these studies can be criticized on the basis of misleading statistical manipulation, and more importantly that the food tested could not be guaranteed to come from genuine organic production. Where food genuinely produced using organic methods has been tested, the results have been much more clear-cut."
—Nicolas Lampkin [10]

This observation is also referred to in the Soil Association's report 'Organic Farming, Food Quality and Human Health', in which, of ninety-nine papers reviewed, seventy-one were shown to be invalid, due to bad science, the limited size of the studies, or their limited duration.[11]

Although there is some evidence that levels of pesticides found in food in the industrialized countries have been steadily decreasing since the 1950s, there is no call for complacency, partly because the levels once considered safe are regularly revised downwards as a result of new research, and partly because of new evidence of pesticide 'cocktails' many times more toxic than single ones on their own.[12] However, more recent and worrying evidence, published by the Pesticide Safety Directorate, has shown that in the UK pesticide residues found in non-organic foods were up in 2002 over 2001 (43 compared with 39 per cent). Unfortunately in many Third World countries the daily intake of pesticides is much higher.[13] Over 86 per cent of the UK public say they do not want any residues in their food. However, Government tests repeatedly show that up to a third of our food contains pesticide residues. The Government's Pesticide Residues Committee report for 2003, for which 4,000 food samples were tested, showed that a quarter had pesticide residues. Of the samples tested in 2003-4, fruit and vegetables were the worst, with a third having traces of chemicals. Of the lettuces sampled, over 50 per cent contained residues and half of these contained multiple residues of between two and five different chemicals. Also, 13 per cent contained residues over the maximum level set by the government. The percentage of samples of pre-packed salads containing residues was even higher.

It needs to be explained here how the maximum acceptable levels of residues are set. The so-called safe limits (MRLs, Minimum Residue Limits) are agreed by testing laboratory rats and mice, and a maximum level is set for each chemical. However, no research is done into the impact of residues on vulnerable groups such as children, or on the multiple effects of several chemicals found together, the so-called 'cocktail effect'.

Pesticides have repercussions for our health, the environment, the poisoning of soil life and the pollution of our water supplies. All these points will be covered, but let's first deal with our food. There are three main areas of research into pesticides in our food:

1. Research on whether there are detectable levels of pesticide present, which ones, and if so, how much;

2. Laboratory research on animals and tissue cultures to find out what symptoms and/or diseases the animals suffer from when exposed to the chemicals involved, what changes occur to the tissue cultures, and the levels of exposure that are harmful and/or deadly;

3. Epidemiological and other studies into the effects of these chemicals on humans that have been exposed to them in their normal lives.

It is very important to distinguish between these three. The fact that there are detectable levels of pesticides in our food does not mean a lot, unless it is shown that the amounts present will harm us. Moreover, showing that a certain chemical can produce diseases in laboratory animals, or changes in tissue cultures when applied at much higher levels than we are normally subjected to, is very useful as a guide when looking out for symptoms in the general population, but it only really becomes a strong argument when backed up by epidemiological studies in the general population. In fact it is only the three together that produces convincing evidence. Needless to say, finding a direct connection between exposure to a chemical and evidence of harmful effects in the human population is the most difficult exercise of all, but there is now a substantial body of evidence which can no longer be dismissed.

The examples below are just a small sample of studies done over many years in many different countries. The first three concern pesticides found on foods, which have often been shown to be above or well above the recommended levels. Some show banned pesticides, and many show several pesticides to be present. These examples are followed by studies done into the harmful effects of these agents, even when consumed in microscopic amounts.

The British Association of Public Analysts:
Surveys of Pesticide Residues in Food, 1983

More than a third of the vegetables and fruit tested contained detectable amounts of pesticide residues. Some, like DDT, were already banned or severely restricted. Others had no clearance from the government, and many were above the official safety limits.

Out of the thirty-two lettuces tested, thirteen showed contamination with Lindane, a persistent organochlorine, which is both a carcinogen (cancer causing) and teratogen (causing birth defects). Three had traces of DDT.

Over 33 per cent of the major British fruits—apples, blackcurrants, cherries, gooseberries, pears, plums, raspberries, strawberries and tomatoes—had detectable residues.

Over 33 per cent of the major British vegetables—beans, beetroot, broccoli, cabbages, carrots, cauliflowers, cucumbers, lettuces, mushrooms, onions, parsnips, peas, potatoes and turnips—had detectable residues.

The British government's response at the time was to say that "In most circumstances, occasional exposure to higher than average levels of pesticide in foodstuff has no public health significance."

The Working Party on Pesticide Residues, 1995 and 1996 [14]

Broccoli No residues in UK samples. Organophosphate (OP) residues in 20 per cent of Spanish samples. Thirty-eight pesticides sought, one found.

Carrots Residues in 75 per cent of samples. Two to four residues in over a third. UK samples had residues in 56 per cent. Twenty-one pesticides sought, eight found.

Celery All except one had residues. 50 per cent had two to five residues. 50 per cent of the UK samples had residues and 90 per cent of imported ones. One hundred and three pesticides were sought, nineteen found.

Courgettes No residues found except in one Spanish sample. Forty-nine sought, one found.

Cucumbers Residues were detected in two UK samples and in 50 per cent of imported ones, and multiple residues (two to four pesticides) were in 25 per cent. Thirty-eight pesticides were sought, and six found.

Leeks No residues were detected, eleven pesticides were sought.

Winter lettuce These are among the worst culprits. To simplify the report on retail samples: In 1995 residues were detected in 75 per cent and multiple residues (two to four) in 50 per cent. Pesticides were present in 80 per cent of UK samples. Ten pesticides were sought, nine found. Two growers were prosecuted. 20 per cent of the samples exceeded the maximum residue limits (MRLs). Six samples contained non-approved pesticides and five contained the organophosphate tolclofos-methyl in excess of 1 mg/kg, way over the limit.

In 1996 residues had risen to 88 per cent, with multiple residues risen to 66 per cent. Pesticides in UK samples has risen to 90 per cent. No non-approved residues were detected, but nearly 12.5 per cent of UK samples still exceeded MRLs. Another test showed much better results, but another on wholesale lettuces showed 90 per cent had pesticides.

Potatoes In 1995 residues were found in 50 per cent of samples and multiple residues in 25 per cent. The majority of residues were in the UK maincrop potatoes. twelve pesticides sought, nine found. In 1996 there was a slight improvement.

Spring greens Thirty-seven pesticides were sought, all in UK samples. None was found.

Sprouted seeds Thirteen pesticides sought, none found.

Tomatoes Residues were found in 50 per cent, and multiple residues in 25 per cent. 40 per cent were found in UK samples and 66 per cent in imported ones. Thirty-six sought, nine found.

Survey of Non-Organic Baby Food Bought in the UK, December 2000

Forty-nine per cent contained pesticide residues at or above 0.01 mg/kg.

Thirty-five per cent contained pesticide residues at or above 0.1 mg/kg (ten times more than the proposed limit).

Fourteen per cent contained multiple pesticide residues at or above 0.3 mg/kg (30 times more than the proposed limit).

Despite these figures, the authors of this report were pleased to say that "compared with a previous study, the traces of pesticides found were less than before and were in very small acceptable amounts." Presumably for grown-ups, the 'acceptable' amounts are even higher.

More recent examples come from the UK Government's Annual Report of the Pesticide Residues Committee, 2003.

Apples 71 per cent of samples contained residues, with 39 per cent containing residues of more than one chemical.

Bread 61 per cent of ordinary bread sampled contained residues, with 12 per cent containing multiple residues.

Cucumbers 24 per cent of samples contained residues, with 4 per cent containing multiple residues.

Grapes 63 per cent of samples contained residues, with 32 per cent having multiple residues.

Lemons (2001 report) There were two surveys. The first found 100 per cent containing residues, with 90 per cent containing multiple residues. The second survey found 93 per cent contained residues, with 81 per cent containing multiple residues.

Pears 58 per cent of samples contained residues, with 22 per cent containing multiple residues.

Potatoes 35 per cent of samples contained residues, with 9 per cent containing multiple residues.

Raspberries 57 per cent of samples contained residues, with 31 per cent containing multiple residues.

Rice 54 per cent of samples contained residues, with 10 per cent containing multiple residues.

Spinach 24 per cent of samples contained residues.

Wine 10 per cent contained residues.

It is important to note here that the UK Government residue testing is very hit-and-miss: only staples such as milk, potatoes and bread are tested annually. There is little consistency: one year it could be peaches, the next farmed fish.

Harmful Effects of Pesticides

"The consumer has become increasingly concerned about the possible dangers from small quantities of food additives and pesticide residues in food. The mounting medical evidence linking additives and pesticide residues to food allergies and increased cancer risks has been the primary stimulant."
—Nicolas Lampkin [15]

Lampkin goes on to cite several studies which provide this evidence. For instance, there has been a consistent rise in the incidence of liver cancers over the years, which some researchers are suggesting is due, in part at least, to ingested pesticides, the liver being the filter for poisons that pass through it. Some 97 per cent of all man-made chemicals found in our bodies have been shown to come from the food we eat.

The US Environmental Protection Agency (EPA) ranks pesticide residues among the top three environmental cancer risks.[16] And an article in the *Daily Express* of 9 June 2001 reported work by Dr Chris Hatton, a consultant haematologist at John Radcliffe Hospital, Oxford, on the effects of herbicides and pesticides on health. He has said there was already evidence that they contribute to cancer deaths. The 5 per cent increase in lymph cancers worldwide, even among young people, is increasingly being linked to herbicides and pesticides. A recent survey showed child cancers have increased by 30 per cent over the last thirty years in the UK. The researchers say that this is due either to viruses or to increases in toxins in the environment and food.

The London Food Commission, 1985

The Commission published the results of a study into the effects of pesticides used in modern conventional agriculture and found on food:

> Forty-nine were possible carcinogens (cancer causing)
> Thirty-two were suspected teratogens (cause birth defects)
> Sixty-one were suspected mutagens (cause genetic damage)
> Ninety were found to be allergens and irritants.

The results of this research were rubbished by the Government: "No chemicals with long-term health effects are allowed on sale in Britain." Since then, however, the government has withdrawn from sale two of the chemicals tested: dinoseb (because of possible birth defects) and ioxynil. Moreover, many of the pesticides used in the UK are banned or severely restricted in one or more countries elsewhere.

The Cocktail Effect

It is very difficult to say what level of contamination will produce harmful effects, but as scientists discover more, there is a tendency to revise the recommended levels of ingestion downwards. There is also an increasing recognition of the cumulative effects over time, as well as the 'cocktail effect', which occurs when there is an interaction between a variety of contaminants.

The cocktail effect has only recently begun to be recognized as a phenomenon, and it is increasingly having to be taken into account when studying the effects of contaminants on the human physiology. In June 1996 the US journal *Science* published a report showing that when two or three pesticides were mixed, the combined effects were 1,600 times more powerful than one on its own.[17] A similar study by a New Orleans group found that the combined effect on the human hormone system of pesticides and polychlorinated biphenyls (PCBs) in microscopic amounts was 1,000 times more powerful than the individual effects.[18] Dr Vyvyan Howard of the Department of Human Anatomy and Cell Biology at Liverpool University has said recently that the cocktail effects of pesticides will have to be taken more seriously in the future. His comments are based on an as yet unpublished study which appears to indicate that combinations of pesticides cause greater damage to nerves than the chemicals individually.[19]

The UK Government has recently recognized this shortfall in knowledge and produced a report entitled 'Risk Assessment of Mixtures of Pesticides and Similar Substances'.[20] The report says there is disquiet about the effect and that there is very little evidence of the occurrence and importance of such cocktails. They have obviously not been reading their own reports describing the occurrence of multiple chemicals found in food samples, or the increasing amount of research described above.

Epidemiological Studies

Lynda Brown describes studies into the harmful effects of pesticides in her book *The Shopper's Guide to Organic Food*:

> "Over a hundred experimental studies have shown that widely used pesticides can alter and suppress normal human immune-system responses. This is now being confirmed by epidemiological studies. A new report from the World Resources Institute reaffirms these concerns and suggests that pesticides may worsen, and increase fatalities from, viral illnesses, particularly in children." [21]

Milk

Residues have been found in milk for some time, and are still being detected. Levels have admittedly been falling over time, but residues have been found in 40 per cent of samples over the last decade or so, with multiple residues in 2 per cent. Indeed, traces of now banned and highly toxic and persistent organochlorines, such as DDT, hexachlorobenzene and dieldrin, are still being found on a regular basis.[22]

Lindane has been banned for a number of years in many countries, but only recently in Britain, and was found regularly in milk samples. It is now banned under EU regulations. Although it breaks down more quickly than other organochlorines, it breaks down into benzene hexachloride (BHC), an insecticide in its own right which has been rejected as too toxic in twenty-eight countries, and which is more persistent and more dangerous than lindane itself.[23] The chances are therefore that it will continue to pollute our food for some time. Deplorable as it is, because organochlorines are so persistent in the environment, even organic milk cannot be guaranteed to be free of these substances.

Banned Pesticides in the Third World
Shockingly, many of the organophosphates and other pesticides that have been banned in most Western countries are still marketed in Third World countries, either openly or on the black market. It has been estimated that 50,000 tonnes of obsolete organophosphates have been stockpiled in African countries alone.

Environmental Pollution

Water pollution
The pollution of drinking water is still a problem in the UK and around the world. The major pesticides found in drinking water are herbicides, with organophosphates common in certain areas where sheep-dipping is practised and the textile industry washes wool.

Reports from the UK Drinking Water Inspectorate continue to show pesticides in drinking water at levels above the maximum permitted concentration (one-tenth of one part per million), but these have declined steadily in recent years. These reports are based on figures supplied under law by the regional water companies, which monitor for over 150 pesticide chemicals. In 1999, 641,335 samples were analysed. The limit was breached on 87 occasions (0.01 per cent of samples) involving 12 chemicals—mostly herbicides. While overall pollution levels in UK drinking water have continued to decline in the last ten years, levels of pesticides and herbicides were still found to have breached the EU Drinking Water Standard. As a result companies have to spend huge amounts of money every year to extract the pesticides and herbicides to reach the standards. For example, between October 2003 and May 2004, the levels of the herbicide IPU in the Cherwell water catchment exceeded the standards on 103 days.

The percentage of samples of drinking water that do not pass the test has declined from 1.3 per cent in 1992 to 0.14 per cent in 2001, but at what financial cost?[24]

The Costs

The capital costs of installing equipment to remove pesticides from water have been put by the water industry at one billion pounds, and the ongoing operating costs at a figure of between tens of millions of pounds up to the low hundreds of millions per year.[25]

The Effects on Birds

The RSPB is pressing for more support for organic farming, and the reason is the effects of pesticides. Seventeen per cent of all wild flowers in the UK grew in cereal fields until the introduction of herbicides. The insects that live off those flowers have declined, and so have the birds that live on them. There has also been the problem of increased infertility in birds in the past due to pesticide contamination—particularly organophosphates such as DDT. As a result of these effects, ten farmland species have declined by 50 per cent or more over the last 25 years, including:

> *Tree Sparrow* 89 per cent
> *Grey Partridge* 82 per cent
> *Bullfinch* 76 per cent
> *Spotted Flycatcher* 73 per cent
> *Song Thrush* 73 per cent
> *Lapwing* 62 per cent
> *Reed Bunting* 61 per cent [26]

Research done by the British Trust for Ornithology in 2004 shows that Song Thrush numbers have increased over the last ten years, but even allowing for this, the overall figures still show a 51 per cent decline over the last 35 years.

Moreover, in 1998 a study by the British Trust for Ornithology found significantly higher populations of soil invertebrates on organic than on conventional farms. This resulted in higher populations of overwintering birds such as rooks, fieldfares, jackdaws and redwings. Fieldfares did particularly badly on the conventional farms, with only a twentieth of the population of the organic systems.[27]

Soil

Soil pollution is probably one of the more insidious, yet unrecognized forms of pollution. As we saw in Chapter 3, among the essential organisms for a healthy soil are the mycorrhizae fungi. These are very sensitive to chemical fertilizers and pesticides (especially, of course, fungicides). Worms are also greatly affected by pesticides, especially insecticides. In fact most of the different forms of life in a living soil are reduced by conventional farming practices.[28] What then fill their place in the ecosystem tend to be disease organisms and microorganisms that enjoy both the high levels of synthetic nutrients and the lack of competition from others. In many soils where the monoculture of cereals is practised year after year, the soil life is almost extinct.

The Costs

The estimated cost of cleaning up our water and removing excessive amounts of pesticides is around £200–221 million per annum. On top of that, a further £799 million–£1 billion has been spent since 1992 on capital investment on the technology to combat pesticides in our water.[29]

There is also the cost of cleaning up nitrates, mainly from nitrogen fertilizers. These additional costs are estimated at around £10 million per annum, with the capital investment required to deal with the problem between £148 million and £200 million.[30]

Professor Jules Pretty, of the Centre for Environment and Society, Essex University, has calculated the costs to the consumer of 'industrial farming', taking into account such factors as removing pesticides, nitrates and phosphates from drinking water, sorting out BSE, dealing with bacterial outbreaks in food and the replacement costs of the annual losses of biodiversity, hedgerows and stone walls. The grand total came to £2.3 billion pounds per annum, or £208 for each hectare of farmed land every year (£84 per acre)![27] As he says, the consumer is paying three times for their food, once at the check-out, again through agricultural subsidies, and finally for cleaning up the pollution caused by modern agricultural practices. This nails the lie of 'cheap food'.

Antibiotics

Antibiotic residues have been found in meats, milk, eggs and farmed fish. Even though the amounts are small, and have no discernible effects on us, there is growing concern about the increasing evidence of the development of so called 'superbugs', in other words bacteria that have mutated to resist a greater and greater range of antibiotics which are also used in human medicine. A WHO meeting of seventy health experts in 1997 concluded that: "Resistant strains of four bacteria that cause disease in humans have been transmitted from animals to humans and shown to have consequences for human health. They are Salmonella, Campylobacter, Enterococci, and E.coli."[31]

In modern agriculture, antibiotics are not just used for emergencies such as infections; since the 1960s they have increasingly been added to the everyday diet of intensively farmed pigs, poultry and fish. Some 437 tonnes of antibiotics were used on UK chicken farms alone in 2001. Despite this, 66 per cent of all chickens grown intensively in the UK are infected with the campylobacter bacteria. About 11,000 tonnes of antibiotics are used in the US every year on animals, of which 80 per cent are purely for growth promotion. In the UK 360 tonnes are used on farm animals each year, of which 80 per cent is not used to cure disease, but to prevent it (prophylactic use) as well as for growth promotion, necessitated by the intensive forms of animal husbandry which are now commonplace.[32]

Without the daily use of antibiotics, intensive livestock rearing would become untenable. This form of farming provides the ideal environment for the development of disease: large buildings packed with one type of animal, often stressed and also kept in warm conditions. When disease breaks out, it spreads very quickly. Despite the routine use of antibiotics in intensive livestock systems, 23 per cent of pigs carried salmonella and 95 per cent carried campylobacter, according to the Meat & Livestock Commission in 2000.

The increasing ineffectiveness of some of the most valuable antibiotics used in human medicine is due partly to overuse by the medical profession, but one of the causes is increasingly being seen as their use in intensive farming. In the past, the effect of intensive use was underestimated, but it is now being taken much more seriously, particularly the widespread use of banned antibiotics which are readily available on the black market. These antibiotics are supposed to be used only in human medicine, precisely to prevent their effectiveness being degraded by overuse in intensive farming. In 1960 only 13 per cent of American human *staphylococcus* infections were resistant to penicillin. By 1988 the figure had risen to 90 per cent, and it is accepted that much of the blame for this is the routine use of antibiotics in intensive farming systems.

When the UK's Food Standards Agency was originally conceived, it was recommended that one of its remits should be to police the use of antibiotics in agriculture and that it should be given the appropriate powers to do so. Unfortunately, as a result of lobbying by the farming industry this role was taken on by DEFRA; in other words the industry insisted that it should police itself.

Nitrates

The problem of excessive nitrate contamination in both our food and water supplies is due almost entirely to the use of nitrogen fertilizers. The emphasis has been on the contamination of water supplies, but while 20 per cent of our daily intake of nitrates comes from water, 70 per cent comes from the vegetables we eat. Lampkin wrote:

> "Nitrates are taken up very readily by crops, and if they are not used immediately in the formation of protein, they are stored in the cells in their original form. There is then the risk that when nitrates are ingested or cooked, they convert to nitrites which can potentially combine with amines to form carcinogenic nitrosamines. Nitrites can also form carcinogenic compounds with certain pesticide residues, such as dithiocarbamates (used as fungicides)."[33]

The Effects of Pollution on Organic Husbandry

Unfortunately, in this polluted day and age, organic food can never claim to be totally 'chemical-free' for a number of reasons. If a farmer has converted from conventional to organic farming there might be pesticide residues in the soil. DDT in particular is extremely persistent, even though its use has been banned in this country for many years. Spray drift from neighbouring farms can also occur, and heavy metals from sewage sludge can also be a problem. All we can say is that the risk of contamination is very much less with organic products.

BSE

The BSE crisis epitomizes the mess we can get into when specialists become detached from reality and common sense. Bovine spongiform encephalopathy (BSE) is a transmissible, neurodegenerative, fatal brain disease of cattle, which was in a sense created by animal nutritionists. There is an almost identical disease in sheep called scrapie. The generally accepted theory of the origin of the BSE outbreak is that as a result of feeding ground-up slaughterhouse remains (including sheep's brains) to cattle, a mutated form of scrapie started to affect them. The disease then jumped the species barrier again, this time to humans who ate affected beef, and was also spread through blood transfusions. This has appeared in the form of a new version of Creutzfeldt-Jakob Disease (CJD), which already existed as a rare disease in humans, and has almost identical symptoms to both scrapie and BSE.

Another possibility is that there was a spontaneous occurrence in cattle, which was passed on by feeding the remains to other cattle. It is now generally accepted that all three versions of the disease are caused by prions, infectious protein molecules which cannot be broken down by the usual means of sterilization and/or drying of the waste meat products.

When the report of an inquiry into BSE was published on 26 October 2000, the discussions in the media revolved around the incompetent handling of the crisis by ministers and civil servants. What was overlooked in the discussions was the assumption that it was all right to feed herbivores animal protein in the first place!

Animal protein has been fed to herbivores for years: there are records going back to 1860 of bone-meal and horn-meal being fed to farm animals, and it has been a regular practice to feed fish-meal to cattle for many years. A neighbour of ours suggested many years ago that we would get better milk production from our cows if we included fish-meal in their rations. I thought at the time that it sounded unnatural, but it was common practice at the time.

Farmers have pleaded ignorance, saying that they did not know what was in the compound mixes they bought, because, until the law was changed recently, manufacturers of animal feeds did not have to state the ingredients on their packets. However, any farmer who has been to agricultural college or read books on animal nutrition—or indeed has made up his own mixes based on common formulas—would have known what was going on. That is precisely why we stopped buying compound foods on our own farm in the early days and made our own mixes up, largely with our own ingredients.

When fish-meal became prohibitively expensive, the nutritional scientists looked around for some other sources of protein and thought it would be all right to use dried and ground waste products from slaughterhouses. This is the insanity of the linear, unconnected thinking that is prevalent amongst specialists today. The 'logic' goes thus: 'The important thing is to squeeze as much milk out of each cow as possible and to grow cattle as fast as possible. Animal protein is more concentrated, therefore it would be more efficiently absorbed and has the added advantage of being cheaper.' When limited intellectual thinking overshadows common sense and sound instincts, this is the kind of stupidity that occurs.

The voluminous report into BSE has failed to point out the most important lesson of all: not just that scientists who were worried about how serious the problem was were not listened to, and were stopped from speaking out or lost their jobs if they did, but that there is something fundamentally flawed in the overspecialized scientific process and thinking, combined with the desire to make money, that led nutritionists to deny that there was anything wrong with feeding the ground-up remains of cattle and other animals to cattle in the first place. It is just this sort of thinking that resulted in recommendations to battery poultry farmers in the past to mix chicken faeces and carcasses with the food for the live hens.

Organic farmers have not fed animal proteins to cattle and sheep since 1983, when the practice was banned by the Soil Association. As a result there have been no recorded cases of BSE on organic farms with closed herds —where all replacement animals were bred on the farm. The only cases recorded have been where cattle have been bought in from non-organic farms, or on recently converted farms where cattle were exposed to contaminated feed before conversion.[34]

The cost of this totally unnecessary fiasco has been astronomical: over £7 billion.[35]

Telling the Wood from the Trees

As Jules Pretty says:

> "Pests and diseases like monocultures and monoscapes because there is an abundance of food and no natural enemies to check their growth. In the end, they have no fear of pesticides, as resistance inevitably develops within populations and spreads rapidly unless farmers are able to keep using new products." [36]

The more monocropping and the less biodiversity there is on our farms, the more pests and diseases will proliferate. The more we farmers grow single crops, without hedges, trees, other crops, animals, weeds and insects, the more Nature fights back. It is we who create the problems. Instead of recognizing where the problem lies, we are convinced that high incidences of disease are a natural phenomenon, and increase our use of pesticides in response.

We need to recognize that there is no battle between us and Nature. We will never eliminate all pests and diseases, but there is growing evidence from around the world that they can be controlled by a more holistic approach.

Chapter 5

Genetically Modified Foods

"So do you not feel that, buried deep within each and every one of us, there is an instinctive, heart-felt awareness that provides—if we allow it to—the most reliable guide as to whether or not our actions are really in the long-term interests of our planet and all the life it supports? This awareness, this wisdom of heart, may be no more than a faint memory of distant harmony, rustling like a breeze through the leaves, yet sufficient to remind us that the Earth is unique and that we have a duty to care for it. Wisdom, empathy and compassion have no place in the empirical world, yet traditional wisdoms would ask 'Without them, are we truly human?' "—HRH Prince Charles [1]

Introduction

'Frankenstein food' scare headlines in the press have detracted from the very important debate that needs to be conducted about GM foods, especially with opinions so entrenched on both sides. Because of this, and also the difficulty of separating the science from commercial interests—the general public are unsure what to think. For many people involved in organic production, however, GM crops seem irrelevant to the challenges that agriculture and society face today. There are genuine and growing concerns about the dangers of genetically modified organisms (GMOs) amongst many members of the scientific community and environmentalists, concerns about the unethical and arrogant stance taken by many of the commercial companies involved in marketing GMOs. But many people involved in the organic field are beginning to feel that they are just an extension of the increasingly discredited and outdated approach to conventional farming. This approach displays more of the same limited and dangerous linear thinking but with one major difference—the results of this new technology are irreversible. Once these altered genes are out in the environment, they cannot be recalled. This is why this technology could be so dangerous.

At the start many people felt that this new technology offered real benefits. But as time went on questions began to be asked. The unprincipled behaviour of most of the companies involved also set alarm bells ringing.

Sceptics are labelled as Luddites by the GMO lobby. But this scepticism is healthy. People are saying, 'There may be benefits, but caution is essential, especially when we also suspect the motives and activities of the companies who promote GMOs.'

The real challenge for agriculture is how can we continue to maintain a vibrant and healthy soil and soil structure so as to provide healthy food on a sustainable basis. Those who work in the organic field cannot see how GMOs will add anything to the creation of true and lasting soil fertility.

We already have the most sophisticated agricultural technology to produce all the food we need, which has proved itself over thousands of years—organic cultivation. With the increased understanding, knowledge and practical techniques gained through the biological and environmental sciences, we have a twenty-first-century agricultural system.

Dreams of Great Promise

Genetic engineering appeared to promise such great advances. From the late 1970s on, those involved in GM research trumpeted the creative uses to which this technology could be put. The possibilities seemed almost endless. However, it is interesting to note that nearly all the most imaginative and beneficial uses that this technology could be put to have so far not been tried. What we have had instead are crops that are resistant to Roundup weedkiller so that the company involved, Monsanto, can sell both the seeds and of course more of its weedkiller; or crops, like GM cotton, with the genes of the bacterium *Bacillus thuringiensis* in them, which will not only kill any caterpillar that eats the leaves, but also any other insect that eats the pollen, such as bees. Then there has been the attempt to sell GM crops, primarily to Third World countries, that could only be grown once because the subsequent seed was sterile, stopping farmers from saving it in the traditional way and forcing them to buy new seed each year.

This appears to most observers to be about making money through patents and cornering the seed market, rather than genuinely helping to improve agriculture and solve world food shortages.

Natural Process?

The main argument that the biotechnology industry regularly trots out is that what they are doing is not new; it is just a further extension of two processes that humans have been using since time immemorial: cross-breeding closely related plants, or animals, and selective breeding. However, these practices mimic the natural genetic changes and processes of natural selection that occur in Nature in all species over time. To argue that genetic engineering is just an extension of these processes is a complete distortion of the truth.[2]

Moreover, the GM companies themselves are contradictory in their claims. One minute they say what they are doing is unique, the next they say that it is perfectly natural. As Vandana Shiva has noted:

> "When property rights to life forms are claimed, it is on the basis of them being new, novel, not occurring in nature. But when it comes time for the 'owners' to take responsibility for the consequences of releasing genetically modified organisms, suddenly the life forms are not new. They are natural, and hence safe."[3]

Genetic engineers are able to do something not possible in Nature. They can take the genetic information from one totally different species, such as a bacterium or a fish, and splice it into the genome of a squash plant, something that could not happen in Nature because there are natural defences against mating with different species. These inhibitions have probably evolved for very good reasons. Even when closely related species mate, as with goats and sheep, the pregnancy does not reach full term, or in the case of mules, the resultant animal is sterile. Does it not occur to the genetic engineers that there might be a natural evolutionary reason for these restrictions, or are they seen as just an irritating evolutionary mistake to be overcome?

Ever since life began, evolution has occurred through a process of sequential unfolding. Each of these small incremental changes in the genes has been rigorously tested by natural selection at every step, and any weaknesses eliminated. To transfer genes from one totally different species to another violates and interrupts the mechanics of evolution itself. This technology is not the same as traditional cross-breeding and selective breeding, whatever its advocates say.

In November 1994 Dr John Fagan, an award-winning researcher into gene regulation and the molecular mechanisms of carcinogenesis, returned a US federal grant of $614,000 in protest, calling for a 50-year moratorium on the environmental release of genetically engineered organisms. He also rejected proposals for further grants worth another $1.25 million. He has said of natural selection and GMOs:

> "Through the spontaneous process of natural selection, nature has been fine-tuning, revising, and polishing the genetic code script of humanity and all other species for an extremely long period of time. To propose that we should revise this text reveals the naive and arrogant assumption that we have grasped that text fully; or that we have at least, understood it sufficiently so that we can see how it can be improved."[4]

Traditional breeding methods imitate Nature's selection processes. Cross-breeding methods imitate Nature's cross-breeding of variations within a species or subspecies but, as in Nature, within narrow parameters. In other words these methods function within the framework of natural laws not outside them—unlike genetic engineering.

Feeding the World

We constantly hear that GMOs have the potential to solve world food shortages, but without the proponents fully explaining how they might do this. One way might be to show that GM crops have higher yields, but the evidence is not there. A two-year study by the US government suggests that, when compared with non-GM crops, they were no better and required the same amount of pesticides. In Kenya, GM sweet potatoes modified to resist a virus were no less resistant than normal varieties and their yields were sometimes lower, according to the Kenyan Agricultural Research Institute. On the other hand, conventional breeding of sweet potatoes in Uganda has produced a high-yielding and resistant variety more quickly and more cheaply.[5]

Our inability to supply everyone in the world with a decent diet has nothing to do with a lack of technology. It has been well documented that where there are local shortages of food, it is nearly always due to war, a failure of administration or political will, unequal distribution of wealth, or drought or other natural disasters. Indeed there is currently more than enough food for everyone on the planet to have an adequate and healthy diet, according to the United Nations World Food Programme.[6]

There is also growing evidence that in Third World countries where organic practices have been adopted recently, there have been examples of increased yields.[7] At the other end of the scale, there has been a disastrous destruction of the soil and environment, resulting in a stagnation in yields, under the so called 'green revolution', which involved huge amounts of fertilizers, pesticides and 'improved' strains of crops being used in places like the Indian Punjab and Haryana and some African countries.[8]

Finally, we must dispel the modern Western myth that single monocrops, grown with chemical fertilizers and pesticides, will always out-yield traditionally produced crops. Many traditional, sustainable small-scale farmers (usually women), have consistently produced more crops per hectare than even the most productive monocropping system, simply because they closely grow several crops at a time on the same plot of land. If one of the crops' yields is measured, it will always underperform modern monocrop systems, but when all the crops are taken into account, these traditional systems far outperform monocrops. Historically, these methods have been the most productive forms of farming on the planet.[9]

Learning More about Less and Less

There is a huge gulf between traditional Third World women farmers, with hundreds of generations of knowledge handed down on how to produce food intensively and sustainably, and research scientists working in laboratories, who have no knowledge of such farmers and their methods. The development of GM crops is yet another example of the technology produced within laboratories by people who know little about farming and have no practical knowledge. In Sir Albert Howard's book *An Agricultural Testament*, he described how agricultural research scientists become completely detached from real farming.

> "The Research Institutes are organized on the basis of the particular science, not on recognized branches of farming. The instrument (science) and the subject (agriculture) at once lose contact. The workers in these institutes confine themselves to some aspect of their specialized field; the investigations soon become departmentalized; the steadying influence of first-hand practical experience is the exception rather than the rule. The reports of these Research Institutes describe the activities of large numbers of workers all busy on the periphery of the subject and all intent on learning more and more about less and less. Looked at in the mass, the most striking feature of these institutions is the fragmentation of the subject into minute units. It is true that attempts are made to coordinate this effort by such devices as the formation of groups and teams, but . . . this rarely succeeds." [10]

Howard's successors, like Jules Pretty, Professor of Environment and Society at the University of Essex, and all the workers out in the field who are practically involved in learning from traditional farmers, are the ones who are really helping to transform agriculture in the Third World for the better. It is they, rather than laboratory scientists, who are being truly instrumental in helping to improve traditional farming practices by combining the best of traditional practice with creative new ideas.

Solving Problems That Do Not Exist

One of the results of research scientists working in laboratories without consulting those in the field is that they try to solve problems that do not exist, or that could be solved much more simply by studies out in the field amongst real farmers. A prime example has been the GM rice called Golden Rice, which contains Vitamin A even when polished. It was developed to combat a form of blindness caused by Vitamin A deficiency, which is prevalent in some parts of the world; as usual, this was done without consulting the farmers and communities involved. Yet women peasant farmers have been growing and encouraging wild and domesticated greens rich in Vitamin A, among other things, with their wheat and other crops. Rather than trying to find a high-tech solution to the problem, why did not the researchers go and ask the peasant farmers in the field, as Sir Albert Howard did in his day, and encourage them to educate others, or to help supply the greens to those who are suffering from Vitamin A deficiency? The answer is not in genetically engineered rice, but in encouraging a more varied diet.

Vandana Shiva has been putting the case for recognizing the true farmers and agricultural experts in the Third World, largely women peasant farmers. In her contribution to the Reith Lectures for 2000 on the BBC under the title 'Sustainability' she said:

> "What the world needs to feed a growing population sustainably is biodiversity intensification, not the chemical intensification or the intensification of genetic engineering. While women and small peasant farmers feed the world through biodiversity, we are repeatedly told that without genetic engineering and globalization of agriculture the world will starve. In spite of all the empirical evidence showing that genetic engineering does not produce more food and in fact leads to a decline, it is constantly promoted as the only alternative available for feeding the hungry. That is why I ask, who feeds the world?

Women in Bengal use more than 150 plants as greens—Hinche sak (*Enhydra fluctuans*), Palang sak (*Spinacea oleracea*), Tak palang (*Rumex vesicarious*) and Lal sak (*Amaranthus gangeticus*)—to name but a few. But the myth of creation presents biotechnologists as the creators of Vitamin A, negating nature's diverse gifts and women's knowledge of how to use this diversity to feed their children and families."

There is an interesting footnote to this GM rice story. Calculations based on the developer's own figures show that an adult would have to eat 9 kg (20 lbs) of Golden Rice to obtain the recommended daily intake of vitamin A.

The Problems

Irreversible Results

One of the most worrying features of genetic engineering is the unpredictable nature of the manipulations involved, and the unpredictable side effects—all of which are irreversible. Of all the problems with GM crops, their irreversibility is the most worrying. Already in countries like the USA, Canada and China, GM genes are out in the environment, where they have been shown to be crossing with both traditional varieties of crops and with closely related wild weeds. There is no way these genetic variations can be withdrawn; they will be there for the rest of time, including any mistakes resulting in the weakening of the gene pool due to the crudeness of the techniques used to produce them. It always has to be remembered that the results are not reversible, which makes this technology uniquely objectionable.[11]

Unforeseen Consequences

There is yet another problem with this technology. Genetic engineering is currently crude and imprecise, which means that other important genes in the cells can be accidentally mutated or inactivated. Scientists do not fully understand what happens when they fuse genes into the DNA of another organism. The slightest change in DNA sequences in the genes, even those that are intended, can lead to an unforeseen chain of events. The information contained in DNA is like a blueprint which has to be activated. The timing of the activation, the sequence, the activation of other processes and the resultant triggering of other sequences is so complicated that no scientist or group of scientists could possibly predict the results. The understanding of these processes is still extremely limited.

As we have already seen, the gradual changes that have taken place throughout evolution have been tried and tested at every stage, with any weaknesses being eliminated or refined over millions of years. Scientists cannot understand the possible repercussions and consequences of each manipulation they do.

Because of the crude method of firing the genes into the cells, it is possible for damage to be done to some other part of the DNA sequence, thereby increasing or decreasing the level of expression of that gene, even if it is not immediately noticeable. These unintended mutations could be introduced into the gene pool, not only of the original species but also neighbouring varieties, including closely related wild species, weakening them again with irreversible results. It is therefore this combination of damage to the gene pool and its irreversibility that is so dangerous.

A large majority of the plants bred from the initial batch of modified cells exhibit faults and have to be discarded, but how many faults pass undetected by the quality testing system? Monsanto's first batch of GM cotton was marketed in 1996. It, and subsequent varieties of GM cotton, had a gene of *Bacillus thuringiensis* inserted to kill two pests, the bollworm and the budworm, which have become increasingly resistant to pesticides and have consequently become serious pests (almost certainly because the same crop has been grown on the same land for many years).

As this first batch matured it became obvious that although the new crop was resistant to budworm it was not killing all the bollworms, so the farmers had to use pesticides anyway. Moreover, many of the cotton buds failed to mature, or dropped off too early. This resulted in many farmers losing yields and profits, and some decided to sue Monsanto. Their attorney, Phillip Maxwell said, "I was in one field where the Bt cotton had grown nine feet high, straight up like a beanstalk with no fruit. It was crazy-looking. Freak cotton."[12]

On 26 March 2002 Monsanto managed to get clearance for commercial planting of three varieties of genetically engineered Bt cotton in India. In three major states it has been wiped out completely, leaving farmers destitute. Not only have new pests and diseases emerged, but the cotton has even failed to prevent the bollworm attack for which it was designed. While it is sold as pest-resistant seed in India, it has proved to be more vulnerable to pests and diseases than the traditional varieties.

Madhya Pradesh, the heart of the cotton-growing belt in India, witnessed the total failure of genetically engineered Bt cotton. The farmers of

Khargoan district are demanding compensation from the company for the failure of their crops, having paid five to six times more for their seed than they would have for conventional seed.

Bt cotton has been afflicted with the leaf curl virus (LCV) in the whole of the northern states of India. Dr Venugopal, the former project co-ordinator of the Central Institute for Cotton Research (CICR), said that while some of the private hybrids and varieties released earlier were resistant to LCV, Bt cotton was found susceptible.

In Maharashtra, which adjoins Madhya Pradesh, the same story has been repeated. In Vidarbha, which is primarily a cotton-growing area, Bt cotton failed miserably. In this district alone, 30,000 ha (74,000 acres) of the first crop failed, completely devastating the already poor farming community. The crop has been badly affected by root-rot disease which it is believed was caused due to the wrong selection of Bt genes, developed for American conditions, but not suitable for India. As a result the farmers of the area are demanding compensation from the Government of Maharashtra, the company Monsanto-Mahyco and the Ministry of Agriculture, for allowing the sale of inadequately tested GM seeds.

In Gujarat there was a heavy infestation of bollworm on the Bt cotton in three districts. Initially it was found to be resistant to bollworms in the early phase of its growth, but as soon as the formation of the cotton bolls had started, the worms attacked them.[13]

It is becoming more and more obvious that this kind of unforeseen problem will continue to happen, especially when the companies involved rush into production as quickly as possible in order to obtain returns on their investments, before testing them thoroughly in smaller field trials in the areas where they are to be used.

Contamination of Traditional Crops and Wild Plants

The contamination of traditional crops and wild species with GMO genes was predicted and has now been confirmed. Contamination has already occurred in the ancient local varieties of maize in Mexico. Recent research by Ignacio Chapela and David Quist of the University of California, Berkeley, has shown that the genes of the traditional maize have been contaminated by GM maize. The effect was strongest along roads, especially main roads, where imported US maize is sold to the locals for their tortillas. The assumption is that the contamination has occurred because the locals sowed the seeds instead of grinding them down for tortilla flour, or alternatively from spilled self-sown seed.

At first the Mexican government, which has banned GM maize, rejected Chapela's and Quist's findings, which were published in *Nature*.[14] Since then it has asked the Director of the Institute of Ecology at the Ministry of the Environment to study samples from sites in two states, Oaxaca and Puebla. A total of 1,876 seedlings were tested and evidence of contamination was found at 95 per cent of the sites, with 8 per cent of the seeds found to be contaminated. The revealing factor was the presence of the cauliflower mosaic virus, which is inserted into GM maize to 'switch on' insecticides that have also been inserted into the maize.

This contamination is made worse by the fact that Mexico is the birthplace of cultivated maize. Many thousands of years ago women farmers in the area started to breed maize from the local wild species, by selection and cross-breeding into the unique varieties they have today. Moreover, in the different climatic and geological areas, local varieties were developed over time that best suited local conditions pertaining and resisted the diseases which were locally prevalent. Here, therefore, is a unique living library which can be used to produce future strains of maize suitable for different climates and with increased disease resistance. If, as is starting to happen, these unique varieties have their genes scrambled by contamination with GM varieties, this living library will be lost for ever.

A disturbing postscript to this story was the pressure put on Ignacio Chapela. One day he was forcibly bundled into a car and pressure put on him to retract his findings. There has also been a smear campaign against him and his findings, on the internet, run by the Bibbings Group, a public relations company, whose clients include some of the top GM companies. The final blow for Chapela came when the university refused to renew his contract, shortly after they accepted a grant from one of the leading biotechnology companies, Novartis.

'Superweeds'

Another danger is the cross-breeding of GM crops with closely related wild species. This phenomenon has already been observed. Many scientists warned that GM plants that had been designed to be herbicide-resistant would cross-pollinate with closely related wild varieties, possibly creating 'superweeds' that would be resistant to the same herbicides that the parent plants were designed to resist. This is exactly what is happening in Canada, as shown by a study done by English Nature, which was published in

February 2002. It has also happened in some areas of the USA. As a result land prices have dropped dramatically in those areas, because nobody wants to buy farms with these new uncontrollable weeds on them.

As we have seen, some GM crops have a gene of *Bacillus thuringiensis* placed in them, which will kill any caterpillar that eats it. The fear is that if these plants cross with wild species, whole new generations of wild plants will become poisonous to insects, including bees, with further damaging effects on wildlife. Rare monarch butterflies have already been observed dying in the USA after eating the pollen from such crops.

Pesticide Use

One of the claims made by the biotechnology industry is that GMOs will allow the reduction and even the elimination of environmentally toxic pesticides, making them environmentally friendly. Although there are examples of decreased use of pesticides on some GM crops, there are many examples where the opposite has been shown to be true. In 1999 an analysis of 8,200 university research trials in the USA showed that farmers who grew Roundup-ready soybeans (50 per cent of the crop grown in 1999), which are genetically engineered to survive spraying with Monsanto's weedkiller, used two to five times as much herbicide as before, presumably because the farmers knew that the plants would not be affected and therefore it was safe to spray more to totally eliminate the weeds.

A more extensive report by Dr Charles Benbrook of the Northwest Science and Environmental Policy Center in Sandpoint, Idaho has shown that a total of 223 million ha (550 million acres) of GM maize, soybeans, and cotton in the United States since 1996 has resulted in an increase of pesticide use of around 22,678 tonnes. The report, published in November 2003, draws on official US Department of Agriculture data on pesticide use by crop and state. Although in the first three years of commercial sales the use of pesticides did drop by 11,520 tonnes, between 2001 and 2003 it increased again by over 33,110 tonnes.[15]

Human Health

As yet there is only circumstantial evidence to suggest that some GM foods have had harmful effects on human health, but there are worrying signs. At Corve University, researchers fed Aventis's T25 GM maize to broiler chickens, although its commercial use was for a silage crop for use in the feeding of cattle. Eight per cent of the chickens died during the trial, which is more than twice the national average of 3.8 per cent.

FlavrSavr tomatoes, produced by the US biotechnology company Calgene, produced lesions in rats. On a danger scale of 1 to 4, the effects were in the range 2 to 3. Nevertheless, the company described these effects as 'mild'. Despite the concerns of US scientists, they were approved in both the USA and the UK, although they are not sold in the UK.

Dr Arpad Pusztai's experiments at the Rowett Institute in the UK found gut lesions in rats following the consumption of GM potatoes with an inserted gene for lectin production. These rats showed no signs of lesions when they ate non-GM potatoes or lectin alone. His research was peer reviewed and published in *The Lancet*. Both the journal and the Royal Society said that the results should be investigated, but after heavy attacks by the biotech companies, his research and similar findings have been ignored.

Although all these effects were observed in rats and chickens and not humans, there is a possibility that some GM foods could be harmful to humans. As soon as Dr Pusztai went public, his findings were rubbished, he was fired, and when he and his wife went abroad for a few days to escape the publicity, his house was burgled. Apart from a bottle of whisky and some foreign currency, all his research papers were stolen. This was followed by a break-in at his old laboratory at the Rowett Institute. I do not usually credit conspiracy theories, but the treatment of people like Ignacio Chapela and Dr Pusztai, disturbed me.

There is also recent evidence that DNA is not broken down as rapidly in the gut as was previously supposed, thus increasing the risk of gene transfer from GM foods. Newcastle University's research on gene transfer, commissioned by the Food Standards Agency, found inserted genes entered gut bacteria in three out of seven human volunteers, and this has implications for many GM foods that contain bacteria marker genes which provide resistance to a range of antibiotics. As a result, there is the possibility of an increase in antibiotic resistance in human gut flora. Following the publication of this research, the British Medical Association stated: "The BMA believes that the use of antibiotic resistant marker genes in GM foodstuffs is a completely unacceptable risk, however slight, to human health."

In the UK there has been a 50 per cent rise in soya allergies since the import of GM soya into the country. Doctors have also reported a rise in Ireland, and in the USA food-derived illnesses are believed to have doubled over the last seven years—coinciding with the introduction of GM food.[16] There is no proven connection, but it is surely worth further investigation.

Genetically Modified Foods

There are fears that edible crops containing genes of *Bacillus thuringiensis* and other 'pesticides' are already interfering with natural human intestinal flora, creating digestive problems in the American population. A recent article in the *Chicago Tribune* reported that the medical damage to the human intestinal tract and to health generally by the 'pesticides' genetically contained in grains has become a silent pandemic.

As yet there have been no epidemiological studies on the effects of GM foods on human health, and the biotech companies continue to claim there have been no ill effects, but the increasing evidence cited above is certainly a cause for genuine concern. Because of the unpredictable results of this crude technology, an increasing number of doctors and others involved in public health are calling for the testing of GMOs to be as rigorous as those on drugs before they are allowed onto the market.

Inevitable Mistakes

The drive to market this technology faster than is sensible, in order to recuperate the companies' investment, is fraught with danger. It is becoming increasingly clear that this technology will follow the same developmental path as atomic power, and chemical fertilizers and pesticides:

1. Basic research leads to new conceptual breakthroughs.
2. Scientists point out the potential applications of this new knowledge, claiming that they promise miraculous benefits.
3. Business and industry hurriedly capitalize on this knew knowledge, implementing applications on a large, commercial scale before scientific evidence regarding the possible dangers is available.
4. Once applied on a large scale, serious side effects become apparent.
5. The companies and scientific advisers to the government deny that there are any problems, and that there is any evidence of harmful effects.
6. Society retrenches, often at great cost, and demands research concerning the dangers of the technology.
7. Research results are then used to develop safer applications of the technology.

This is the pattern that has been repeated over and over again in the nuclear and chemical industries, and it is likely to be repeated again with GMOs. The effects of GMOs cannot be reversed or corrected and will afflict all future generations, and therefore the last stage becomes very difficult, if not impossible, to implement.

One Vast Experiment

What we are really talking about here is one vast experiment in the countryside, with the environment, animals, and us humans as the guinea pigs. When this technology was very new, there were few restrictions on it, so the companies involved moved forward as fast as possible before the inevitable regulations came into force. They encouraged farmers to grow as many GM crops as possible as quickly as possible, pushing ahead with little regard for the possible consequences.

This strategy worked very well in the USA and some other countries, but singularly failed in Europe and India among others. In the USA it went well partly because it was the first country to apply this technology commercially on a large scale, and partly because there is still a widely held belief there that anything that is new and scientific must be good. However, there has been considerable surprise in the USA at the backlash in Europe, which has resulted in a belated movement against GMOs. A recent poll by Reuters found that 54.4 per cent of Americans were concerned about GMOs, and 33.3 per cent said that farmers should stop growing them.

In Europe, on the other hand, after the BSE scares, there has been a much more sceptical response to the lavish claims from the GM lobby. The more the 'experts' say 'Trust us', the more the general public do not. The power of the buying public in the UK, for instance, has led to the supermarkets withdrawing GMO foods from their shelves and using non-GM ingredients in their own-brand items.

In countries like India, the reasons for the backlash have been different. The attempt to push GM crops onto the farmers, encouraging them to grow crops whose seeds are so expensive that even where there are small increases in yields their profits are greatly reduced, and where the traditional saving of seeds to sow next year has been stopped by private detective agencies employed by the biotechnology companies, smacks of old-style imperialism dressed up in modern clothes. In 2001, thousands of hectares of Bt cotton were illegally sown in Gujarat. The seed was sold to local farmers as being bollworm-resistant, but without telling them it was GM cotton. In March 2002 the Indian government approved the commercial cultivation of Bt cotton in India. Only one company is licensed to sell Bt cotton: Mahyco, in which Monsanto holds 26 per cent of the shares.

The Politics

It is not enough to concentrate on the pros and cons of the technology itself, however, without taking into account the practices of the companies involved and their supporters. Many critics feel deeply concerned about this unique combination of doubtful technology and the cavalier way it is being marketed, especially in Third World countries.

The Companies' Tactics

When people woke up to what was happening, and the inevitable backlash came, the war between the companies and their critics started to get very dirty. With billion-dollar investments and years of development at stake, the companies involved were not going to wait around for their growing band of critics to get the upper hand. Their strategy has been to attack on as many fronts as possible at the same time. They have spent huge amounts of money employing public relations people to devise a global promotion programme to win over politicians, the media, scientists, and of course the public, and to discredit their opponents.

They employ well known PR companies to produce propaganda in the form of an educational campaign. They seem to believe that the public are ignorant and irrational, influenced by environmentalists, who have a built-in tendency to 'Luddite technophobia', and that if they were only supplied with the facts about GMOs they would come to understand what a wonderful technology it is. To this end they enlist supporters such as officials of the seed companies and so-called independent experts to make positive claims. Farmers have been encouraged to trumpet the benefits of biotechnology, because the PR companies know that the farmers are trusted by American consumers.

Another approach has been to win over as many influential people with the ear of the government as possible, and to make sure there are as many sympathetic scientific advisers inside the government itself. For example insiders in the Food and Drug Administration (FDA) in the USA have told how top-level officials leave to work for Monsanto and Monsanto executives move to the FDA, in what they call 'the revolving door'. Biotech companies have also supplied advisers on GMO technology to many governments around the world.

At the heart of organizations like the US Biotechnology Industry Association, are behind-the-scenes efforts to 'educate' journalists. Tom

Hoban, a biotech proponent, has observed that these efforts are important because media stories will be crucial to shaping public opinion. The companies also use less 'benign' methods. A worrying trend in the USA recently has been the Agricultural Product Disparagement Laws that the industry has had enacted in more than a dozen states. These laws make it much easier to sue writers and journalists if they have written disparaging things about the food and biotechnology industry. This is an appalling trend in a country that prides itself on being the bastion of free speech.

In the USA and Canada, farmers have also been prosecuted for saving GM seeds for re-seeding, often when they have not done so. In countries like India, there is a tradition of saving seeds for resowing and distributing to their neighbours. Many of the Indian farmers, partly through ignorance and partly because of traditional practices, saved the seeds. Private investigators have been employed to gain evidence for the prosecution of these farmers for saving the seeds, much to their confusion and distress.

UK Parallels

Similar connections between the industry and the government exist within the UK. The *Daily Mail* reported in July 2003 that in the first two years of the Labour government, GM companies met with government officials eighty-one times, and that Lord Sainsbury, the largest single donor to the Labour party (around £9 million), is a strong advocate for GM companies—indeed, until he joined the government he had substantial shares in Diatech Ltd, a biotech company which owns worldwide patent rights over a key gene used in the process of GM crops.

As Under-Secretary of State for Science, Lord Sainsbury is responsible for the Office of Science and Technology in the Department of Trade and Industry. Since becoming a member of the government, his assets are independently controlled by a blind trust under the terms of the ministerial code of conduct and he is not allowed to sit on any committees that are discussing GMOs or the biotech industry, but he does have the ear of the Prime Minister, who is a strong proponent of GM technology and has attacked what he terms "GM luddites". He has championed the biosciences in several speeches, and sees their development as important for the economic health of the UK.

Owing to public disquiet about GMOs, the UK government instigated limited field trials and provided a small budget to encourage a formalized public debate about GMOs called 'GM Nation?'.[17] The results of the formal debates were 5 to 1 against GMOs, but obviously, the audiences at these meetings were to a certain extent self-selected. However, the organizers also ran a number of meetings of people chosen at random from the population, many of whom had no views either way. After hearing both sides of the arguments, and after question and answer sessions with experts, they came out 4 to 1 against.

The big problem for the biotech industry in Europe, and the UK in particular, is that the vast majority of the population do not want to buy GM food, and the shops and supermarkets responding to their customers, won't stock them. Defenders of GM may consider those who criticize GMOs to be 'technophobic luddites', but there is an increasingly informed public which is not taken in by the hype. Whether the authorities are listening, however, is doubtful.

Regulatory Authorities

The regulatory authority in the UK, whose job it is to monitor and advise on GMOs and their use, is the Advisory Committee on Releases to the Environment (ACRE), chaired by Lord Grey. However, the effectiveness of this committee has been seriously called into question recently, because of its recommendation to allow the commercial use of Aventis's GM maize T25. Dr Stephen Keston, a senior researcher at Bristol Veterinary School, who has expertise in doing this sort of trial, has said these results would have automatically prompted them to do a larger trial, or reconstruct the trial and do it again. He went on to say that the study was not constructed well enough, and therefore could not provide any useful information. His conclusion was that this study was nowhere near adequate to support the permission for Aventis to market T25.[18]

ACRE had asked for advice on the study from the 'Advisory Committee on Agricultural Feeds', who said they could see no reason why T25 maize should not be fed to cattle, having looked at the test results.

There is obviously a serious loophole in the existing arrangements. ACRE have said they have doubts about the trials themselves, yet they advised the approval of T25 maize, which is already being grown commercially in our fields. Although it has only been approved for cattle feed, there is every reason to suppose that similar problems could arise with GM foods designed for human consumption.

Dr Stephen Keston argues that GMOs should be tested as rigorously as veterinary medicines, which are tested in extensive trials to GLP (Good Laboratory Practice) standards and have to be reported in a strictly prescribed fashion. The GLP standards are international standards which apply to a wide range of laboratory practices, including the production of veterinary medicines and pharmaceuticals in general. Dr Keston says this is the kind of quality and quantity of testing that he would like to see for GMO plants which have the capacity to do quite a lot of damage to quite a lot of systems.

Pressure on Farmers

In countries like India, needless to say, the practice of prosecuting farmers for saving the seeds of their GM crops, as they have done with non-GM seeds traditionally for hundreds of generations, has been a PR disaster. This kind of corporate bullying has not gone down well. However, the practice is not restricted to Third World countries, as it is also occurring in the USA, with company representatives threatening farmers suspected of saving and resowing seed from their crops. One example was Rodney Nelson, from North Dakota a conventional farmer, with 3,600 ha (9,000 acres) of land, who in 1998 grew Roundup-ready GM soya on 24 ha (60 acres).[19]

He was disappointed with the yields, but was persuaded by his dealer to grow 600 ha (1,500 acres) in 1999. He paid Monsanto $56,240 for the seeds plus $18,800 as a 'technology fee'. Once again the crop yielded poorly, and investigators from Monsanto began to suspect that he was saving seed from his crop. They filed a law suit, which meant his family nearly lost everything. As he said:

> "It was a hellish ordeal with Monsanto, two years of hell. And this is not an isolated case—hundreds of farmers throughout America have been facing the same situation. GM crops have nothing to do with feeding the world. This is all about patents and the money patents make."[20]

However, despite all their efforts, there are hopeful signs that more of the American public are increasingly worried about GMOs. Even the US National Farmers' Union is now urging restrictions on them. It has called for a moratorium on the licensing of new transgenic animals and plants, and issued a list of precautionary measures they would like to see introduced. It says it "recognizes GMOs have created a series of ethical, environmental, food safety, legal, market, and structural issues that impact everyone in the food chain."

Pressure on the EU

The EU has unfortunately given way to US pressure to begin to lift its moratorium on GM foods. EU leaders have tried to balance this with new regulations that make the labelling of such foods mandatory, so that consumers at least have a choice of what they eat. However, the US has responded that this labelling as an illegal restraint on trade. It is challenging this approach at the World Trade Organization (WTO).

Misrepresentation

On 9 July 2003 an extraordinary talk was given by Steven Druker, a US public interest attorney, exposing the ways in which the US and EU administrations have been handling the GM issue. He showed how the US attack against EU policy on GM foods is based on a systematic deception by the US administration. He also revealed how the EU regulators are themselves misleading the public about GM food and have never truly upheld the 'precautionary principle' that they were supposed to be signed up to.[21]

Druker is the founder of the Alliance for Bio-Integrity, a non-profit organization dedicated to promoting technologies that foster human and environmental health and addressing the problems of those that do not. In this role he has participated in scientific conferences around the world, including Israel and Denmark, has made presentations in the USA, Canada, New Zealand, Australia and sixteen countries in Europe, Asia and South America. He was a featured speaker at a meeting of MPs at the House of Commons, and was invited to the White House to discuss the environmental risks of agricultural biotechnology at a special meeting.

In his talk in Brussels, he explained how the FDA covers up the warnings of its own scientists about the risks of GM foods and systematically misrepresents the key facts. He explained how GM foods had originally been allowed on the market in the United States, Europe and other countries as a result of similar misrepresentations by the US government, and how their continued marketing depends on the continuation of these misrepresentations.

This fraud was exposed by a lawsuit that Druker organized against the FDA, which forced the agency to divulge its internal files on GM foods. They revealed how the politically appointed administrators had covered up the extensive warnings of their own scientists about the unique hazards of these foods. In organizing the suit, he assembled an unprecedented coalition of eminent scientists and religious leaders to stand as plaintiffs.

He went on to show how the FDA's records revealed that its own scientific experts overwhelmingly concluded that genetic engineering differs from conventional breeding and has the unique potential to produce unintended and essentially unpredictable new toxins and other harmful substances. These experts warned that a GM food could not be considered safe unless it had undergone rigorous toxicological tests using the whole food. There was a uniformity of opinion that "the process of genetic engineering and traditional breeding are different, and according to the technical experts in the agency, they lead to different risks."

The FDA bureaucrats, who admit that they have been operating under a White House directive to foster the biotech industry, disregarded this input, and in May 1992 instituted a policy that presumes GM foods are as safe as conventional ones and do not require any testing. Indeed they declared that they were unaware of any information showing that GM foods differed from other foods. They additionally claimed that there was overwhelming recognition among experts that GM foods are as safe as others—which was patently untrue. This duplicity was exposed by a letter sent by the FDA Biotechnology Coordinator to a Canadian health official on 23 October 1991 in which he commented on a document that discussed GM foods, saying: "As I know you are aware, there are a number of specific issues addressed in the document for which a scientific consensus does not exist currently, especially the need for specific toxicology tests."

Steven Druker went on to explain how over nine years the FDA has continued to issue deceptive statements, repeating the mantra that "FDA's scientific review continues to show that all bioengineered foods sold here in the United States today are as safe as their non-bioengineered counterparts." This is despite the fact that at the same time they had "not found it necessary to conduct comprehensive scientific reviews of foods derived from bioengineered plants." The FDA continues to claim there is an overwhelming consensus among experts that GM foods are safe, despite the fact that it has repeatedly been notified through formal channels that numerous eminent scientists consider these new foods to be inherently more hazardous than their conventional counterparts.

Steven Druker went on to show how the EU regulators have ignored the unique capability of GM foods for unpredictable results and have failed to require the kinds of tests recommended by the FDA experts. Instead they have relied on tests that do not adequately screen for the potential negative effects.

The basic approach on which EU and Canadian regulation of GM foods is based—the concept of 'substantial equivalence'—is so unsound that a report from the Royal Society of Canada issued 5 February 2001 criticizes it as "scientifically unjustifiable".

Even within the narrow parameters in which they have operated, the EU regulators have failed to uphold adequate standards. For one thing, the research on which they have relied is generally deficient and does not meet basic scientific protocols.

Druker went on to say that although consumer resistance has prevented most GM foods from being marketed for humans in Europe, large quantities have been imported for animal feed and EU citizens have for years been unknowingly consuming meat, milk and eggs from animals raised on this feed because no labels have been required; nor are they required on meat, milk and eggs produced from these animals. It is interesting to note that the FDA experts themselves have stated that feeding GM crops to farm animals presents "unique . . . food safety concerns" because residues of unexpected substances could make meat and milk products harmful to humans.

Consequently, the marketing of GM foods in the EU is contrary to the guiding principle of the EU's own food law, the precautionary principle, as exemplified by the European Commission's Green Paper 'General Principles of Food Law in the EU, 30 April 1997', which states that it "will be guided in its risk analysis by the precautionary principle in cases where the scientific basis is insufficient, or some uncertainty exists."

Druker claimed that recent evidence provides added justification for concern about unexpected harmful side effects. First, the discoveries of the Human Genome Project released in February 2001 confirm that the foundational assumptions of genetic engineering are overly simplistic and seriously unsound, and indicate that recombinant DNA techniques entail greater potential for unpredictable hazards than was previously suspected even by experts advocating a more precautionary approach.

Secondly, there is mounting evidence of GM plants with substantial and unexpected alterations in chemical composition. Aventis's own data show statistically significant differences between T_{25} herbicide-resistant maize and its conventional counterpart in terms of carbohydrate, amino-acid and fatty-acid composition.

Moreover, the Public Health Association of Australia (PHAA) also analysed Monsanto's data from controlled studies on three of its GM plants (herbicide-resistant maize and canola, and pesticide-producing corn), and in all three cases discovered several statistically significant differences in chemical composition (including amino-acid profiles) between the GM organism and its non-GM counterpart. The PHAA report (October 2000) states that the differences in the amino acids cannot be attributed solely to the known products of the inserted genes and cautions that these plants may contain unexpected—and to date unidentified—new proteins that could be harmful to humans. Recent investigations by Japanese scientists also reveal that Monsanto's data on its Roundup-ready soybean, the most widely planted GM crop, show important differences between the GM bean and its conventional counterpart. For instance, after heat processing of both the GM and non-GM beans, the concentrations of three harmful substances were significantly higher in the GM samples.

Finally, research at the UK's John Innes Centre confirms that the viral promoter used in almost all GM plants can facilitate various abnormal genetic recombinations, which could lead to serious disruptions or to the generation of new and hazardous chemicals. Additionally, experts warn that parts of existing viruses could recombine into novel and more dangerous viruses.

As a result of these and other pieces of research, hundreds of scientists have signed an open letter to the world's governments warning of the hazards and calling for a moratorium on all GM foods. Signatories include professors of biology from Harvard and the Massachusetts Institute of Technology, and the director of the renowned Woods Hole Research Center. Further, nine scientific experts were so concerned that they took the unprecedented step of joining as plaintiffs in the Alliance for Bio-Integrity's lawsuit against the FDA. By asserting that they believe GM foods entail unique risks they refuted the FDA's claim that scientific experts overwhelmingly recognize them as safe. They include a professor of molecular biology at the University of California, Berkeley, the Co-Director of Targeted Mutagenics at Northwestern University Medical School, and a renowned expert in plant genetics at the University of Minnesota, whose declaration to the court stated: "There are scientifically justified concerns about the safety of genetically engineered foods, and some of them could be quite dangerous."

Druker also believes that there is insufficient evidence to demonstrate that any GM food is safe, and that there is a significant degree of doubt about their safety on the part of numerous experts. Accordingly, it is difficult to see how the presence of any GM food on the EU market is consistent with the precautionary principle that EU regulators are legally bound to uphold.

Druker concluded his talk by saying:

> "It's astounding that this venture to radically transform the production of our food continues to be portrayed as based in sound science and rigorous regulation, when in fact it depends on flagrant disregard of both science and food safety laws. Such deception ranks as one of the greatest frauds ever perpetrated. . . . It should be clear by now that genetic engineering is the wrong way for agriculture to go. The technology is inherently risky, but industry refuses to perform the kind of tests that are minimally required to screen for the full range of hazards. And if regulators actually required the tests that sound science dictates, it would be an economic disaster for the industry because, in contrast to the pharmaceutical industry, the food industry is not structured to bear the costs of rigorous long-term toxicological tests for each new product. The economic realities of GM food are incompatible with the scientific necessities."

The obvious choice is to stop ploughing massive resources into genetic engineering and, for a small fraction of the cost, foster the development of safe and sustainable organic agriculture.

Patenting of Traditional Crops

There is yet another side to this story that is equally worrying. Because of the unique form of patent laws in the USA and the Trade Related Intellectual Property Rights Agreement of the WTO, it has been possible for certain biotech companies to take out patents on traditional crop varieties and herbs grown in Third World countries. This is just a form of pirating and stealing. As Vandana Shiva said in her contribution to the BBC's Reith Lectures in 2000:

> "Patents and intellectual property rights are supposed to be granted for novel inventions. But patents are being claimed for rice varieties such as basmati for which my valley—where I was born—is famous, or pesticides derived from Neem, which our mothers and grandmothers have been using."

Rice Tec, a US-based company, has been granted Patent No. 5,663,484 for basmati rice lines and grains.

Basmati, neem, pepper, bitter gourd, turmeric—every aspect of the innovation embodied in indigenous food and medicinal systems is now being pirated and patented. The knowledge of the poor is being converted into the property of global corporations, creating a situation where the poor will have to pay for the seeds and medicines they have evolved and have used to meet their own needs for nutrition and health care.

Such false claims to creation are now the global norm, with the Trade Related Intellectual Property Rights Agreement of the World Trade Organization forcing countries to introduce regimes that allow patenting of life forms and indigenous knowledge.

Instead of recognizing that commercial interests build on nature and on the contribution of other cultures, global law has enshrined the patriarchal myth of creation to create new property rights to life forms, just as colonialism used the myth of discovery as the basis of the takeover of the land of others as colonies.

By a combination of promoting patented GM seeds and trying to patent as many of the existing traditional varieties as possible, these companies are trying to develop complete control over the market.

The Organic Response

For all the above reasons, the organic movement has banned GMOs from organic practices at all stages of production and food preparation. The only way to be sure at the moment that you are buying products without GMOs is to buy organic. Whether it will continue to be possible to maintain uncontaminated crops in the future, however, is in doubt. There are increasing examples of GM contamination affecting the livelihoods of organic farmers in the USA and Canada, including soya and rape seed.[22]

Arnold Taylor, a Canadian organic farmer, says that virtually no organic oilseed rape can be grown in Saskatchewan any more, because the stock is polluted by genes from Roundup-ready GM rape. Terasa Podell, a North Dakotan organic farmer, grows hard red spring wheat, but is worried by the secret growing of GM red wheat in her district. She says to UK organic farmers: "GM is like a cancer that has no boundaries. It's like giving farmers a death sentence. Our organic producers are living a nightmare—please don't allow our nightmare to become yours."[23]

Summary

To summarize, let's sort out the wood from the trees and concentrate on the most important points.

1. The most important thing of all about GMO technology—when applied to crops—is that the changes to the genes are irreversible. Once they are out of the laboratory in the countryside, they cannot be recalled or reversed, unlike all other previous technological innovations. The most worrying thing about this technology is that any mistakes that occur will be permanent.

2. The combination of the crudeness of this technology, and the huge and unpredictable chain of events produced by the manipulation of genes, means that the scientists and technicians cannot be entirely in control of the results.

3. So far, this technology has shown itself to add nothing to the most important challenges that agriculture and horticulture have to face up to in the twenty-first century—namely, how to create and maintain a productive and sustainable healthy farming system that can provide all the food we will require *ad infinitum* into the future. In other words, in the minds of many, GM food has so far shown itself to be a complete and dangerous irrelevance.

Having taken into account all these points, there is still the very real problem of the unscrupulous behaviour of the GM companies—already discussed in this chapter. As the results of the GM Nation public debates showed, this was one of the biggest concerns of many of the public who attended these forums. The way the GM companies have carried out their affairs—often encouraged by governments who are intent on increasing their countries' GNP and are worried about being left behind in the race—has added another dimension to the technical problems already discussed. All the dangers that GM crops pose are exaggerated hugely when the companies involved are determined to promote their products at any cost. The result has been to make an already hazardous situation highly dangerous.

Chapter 6

Maintaining Soil Fertility

"The forest manures itself. It makes its own humus and supplies itself with minerals. If we watch a piece of woodland we find that a gentle accumulation of mixed vegetable and animal residues is constantly taking place on the ground and that these wastes are being converted by fungi and bacteria to humus."—Sir Albert Howard [1]

Nature recycles everything; almost nothing is wasted. When a plant, an animal or even a tree dies, it is consumed by animals, insects, bacteria and fungi. In the process its constituent parts are disassembled to provide food for new life. This, then, is the model to follow in our own human practices.

In this book we are concerned with the recycling of organic material that can be composted and converted into humus to feed back to farms for the production of food, but organic recycling has to be part of a complete programme of recycling on all levels.

More organic recycling is essential, but will only be achieved if a major lead is taken by governments. In the UK, the government is starting to show signs of moving in this direction. Unfortunately, EU and US efforts to reduce landfill have led to the construction of electrical plants that burn a variety of waste, including organic material, to generate electricity, causing pollution and wasting huge amounts of a very valuable resource. There should instead be much more imaginative uses of organic waste.

In this chapter we will look at the principles involved in the manufacture of high-grade compost, from the smallest enterprise to the largest, as well as other ways of utilizing waste organic matter and the most appropriate methods of recycling organic waste from diverse sources, plus the uses of the compost to maintain soil fertility, and how the use of green manures and zero-tillage tie into this equation.

The Theory of Composting

Why is so much emphasis placed on composting in organic circles? Why not spread fresh manure on the fields, as most conventional farmers with livestock do? Why go to the bother of composting it first? These questions are perfectly understandable, and these are the answers:

1. The organic goal is to feed the soil, and more importantly the life within it, not the plant. This is achieved by providing as much humus as possible to encourage a vibrant living soil.

2. To make humus requires lots of high-cellulose plant material such as straw, crop wastes, hedge trimmings etc.

3. To break down this high-carbon cellulose material requires high-nitrogen animal manure and urine, or soft green plant material.

4. In this approach, high-nitrogen manure is no longer seen as a plant food as such, but as a precious activator to enable the creation of a much larger amount of an even more valuable end product—compost.

Sir Albert Howard wanted to understand the biology of ancient Eastern composting methods at the experimental agricultural station at which he worked at Indore in India, between 1924 and 1931.[2] By studying the fungal and bacterial activity at different stages of organic fermentation, and the temperature changes associated with these changes and other factors, Howard was able to come to a thorough understanding of the biological and chemical processes involved in humus production. He knew that in Nature, when there are healthy aerobic conditions, all organic matter rots down to humus with the same stable carbon to nitrogen ratio of around 10 to 1. He also discovered that whatever the proportion of carbon and nitrogen in the original mix, it always ended up in approximately the same proportions.[3] This is because when high-nitrogen manure (waste protein) is left to rot down, the excess nitrogen is given off in the form of ammonia (NH_4) or gaseous nitrogen and wasted. If, on the other hand, there is too much carboniferous material (cellulose), it will take a long time to break down.

To achieve the correct balance of different components, he found that the ideal proportions of the original materials should be about thirty-three parts of carbon to one part of nitrogen.[4] This was one of Howard's master strokes, often misunderstood by those who make compost, on both a commercial and a garden scale. A carbon to nitrogen ratio of 33 to 1 deliberately errs on the side of excessive carbon because in the second stage of composting, starting about five weeks after making a heap, the bacteria start fixing nitrogen from the atmosphere, and under favourable circumstances one can actually gain up to 25 per cent extra nitrogen.[5] Purposely tipping the balance in favour of carbon, in other words having a slight excess of carbonaceous material in the compost heap, means that the extra nitrogen has to be obtained from somewhere else. So the second important factor, is to make sure there is a good supply of air in the compost heap, at least in the early stages. Where there is plenty of air, there is both the necessary oxygen and nitrogen, so the Azotobacter and other aerobic bacteria take the excess nitrogen they need from the air itself. This ensures the final compost ends up with more nitrogen than the original contents—and this is the beauty of the process.

This technique is taken a stage further in municipal composting and on many organic farms, where the compost is frequently turned. Because of a lack of understanding of this basic principle, most composted waste on farms is too high in manure, and as a result there are regular losses of up to 30 per cent of the original nitrogen. Even when fresh manure is spread directly onto soil, there can be losses of between 10 and 30 per cent.[6]

In practical terms, Howard's own formula of a 33 to 1 carbon to nitrogen ratio in the initial mix works out as three parts of vegetable waste (town waste, prunings, straw, horticultural waste, hedge trimmings, crop residues etc., which are high in cellulose and some lignins, and are therefore high in carbon), and one part of animal waste (manure and urine), with varying proportions of grass and clover clipping, young nettles and other young green leaves etc., which are high in protein and therefore largely consist of nitrogen. To encourage the essential fungal growth of the early stage and a vigorous growth of Azotobacter bacteria, one needs to ensure both good supplies of air and the maintenance of neutral or alkaline conditions in the heaps. To help create these conditions, ground limestone (and sometimes wood ash) is added, which keeps the conditions sweet and neutral, providing the ideal environment for the aerobic bacteria, the mycorrhizae and the manure or brandling worms (*Eisenia foetida*), which are so essential for successful composting to thrive. Aerobic decomposition of organic matter is always more efficient than anaerobic. Anaerobic and acid decomposition is wasteful because precious nitrogen is lost to the atmosphere, mostly in the form of ammonia.[7]

Another important reason for composting is that nitrogen, phosphate and potassium are 'fixed' during the composting process. Nitrogen, in the form of positively charged ammonium ions, becomes attracted and attached to organic matter as the compost matures. Moreover, much of the nitrogen becomes a living protein in the form of the bodies of bacteria. The highly unstable water-soluble potassium salts in wood ashes, for instance, are converted into more stable forms in the heap.[8] As we saw in Chapter 3, this results in the steady release of nutrients from the resultant humus as and when the plants need it, rather than a huge wastage of nutrients being washed out of the soil or oxidizing before being absorbed by the plants. To help these processes along, it is useful to have fresh soil in thin layers within the heap and on the surface as a casing. Not only does the soil supply bacteria and fungi to 'seed' the heaps, but clay particles and humus in it also help to bind free ammonia which might otherwise be lost to the air. As we saw in Chapter 2, the knowledgeable farmers of China used to collect large amounts of fresh soil from the fields to mix with their compost heaps.[9]

One of the things that surprises many people is that, if properly made, compost always ends up as a sweet-smelling, crumbly product, however smelly and objectionable some of the original fresh contents were. We have evolved the ability to know instinctively when something is healthy or unhealthy, mostly by smell. Trust this instinct; a healthy vibrant soil or well rotted compost smells good because it is good.

Composting on Different Scales

Household and Garden Composting

Even with the smallest garden you can do your bit to recycle your food and garden waste. Many local authorities now provide bins for recycling organic waste.

There are two ways in which you can compost your waste: in a compost heap or with a worm compost bin. Many local authorities have schemes to promote home composting, including supplying composting units—usually at a discount, or free.

If you have the space, you can build a compost heap, preferably retained by wooden slats. In a small garden you can use the plastic bin type, which you can buy from your local garden centre. The standard compost heap takes bulk waste plant materials, grass mowings, dried seaweed, animal manure in small amounts when available (various dried forms can now be bought at garden centres, including dried blood), kitchen waste, plus the ashes of woody material and pernicious weeds from a bonfire site.

As an alternative—or in addition—one can have a plastic dustbin in a greenhouse or outhouse, with a thriving population of manure worms (brandlings) which feed on kitchen waste—uneaten food, waste meat, vegetable peelings and trimmings, tea-bags, coffee grounds etc. Alternatively you can buy ready-made wormery kits, comprising a bin, worms, bedding material and calcified seaweed, plus instructions. These have the advantage of being vermin-proof.

Although we have four large traditional compost heaps we also have three plastic dustbin wormeries for converting kitchen waste, plus one plastic rain barrel full of water for breaking down and converting persistent perennial weed roots, such as couch grass, dandelion and dock. The worm compost ends up looking like black peat, and after sieving it to get out the coarser material, it is ideal for making up potting compost or as a concentrated soil conditioner which encourages soil life. The covered water-filled rain barrel for perennial weed roots is a Roman idea. The roots rot down and produce a rich liquid manure for potatoes or as an activator for the compost heap from the bottom tap. When they are finally dead, they go onto the compost heap.

Community Composting

Although community composting forms only a small percentage of the total (approximately 0.194 per cent in the UK), it is still a healthy trend that will hopefully grow. It can involve bulk composting of household waste, to be used later by the householders that contribute. Alternatively, it may be a group of allotment holders who collect kerbside or municipal waste, such as leaves collected locally by contractors. Whatever form it takes, it is an increasingly valuable part of the overall recycling of organic waste. The great advantage of such schemes is that the material is not transported very far, making them much more ecologically sound. The material is collected locally, little or no machinery is used in its manufacture, and the final product is used locally.

Farm Composting

Newman Turner was convinced that he did not need to turn the large compost heaps he made on his Somerset farm in the 1950s, because he built them on piles of hedge clippings and other coarse material, and made them lightly, thus providing as much air as possible and delaying compaction.[10] However, with the advances in farm machinery since that time, it is much easier to turn large heaps at least once during the composting process. There is even specialist machinery for inverting compost windrows.[11]

Getting enough high cellulose waste is always a challenge on a farm. One of the problems with most modern varieties of cereals is their short straw. They were bred to withstand nitrogen fertilizers, which forced the older varieties to grow too tall and thin, which made them vulnerable to wind and rain. Some of the older varieties with longer, stronger straw, are better for organic conditions, because the farmer is looking for two crops—the grain and the straw. Straw is the most obvious high-cellulose product; it tends to be used copiously on organic farms for animal bedding when it is available. This material ensures a higher ratio of carbon to nitrogen in the composting process.

However, more can be done to secure high-fibre material. I have long thought there must be ways of saving the hedge and verge clippings from mechanical hedge-flayers by adapting the machinery. On many farms this would provide a substantial amount of useful material. Another possibility is for farmers to look out for waste organic material from a number of local sources for incorporation into the compost heaps.

Research in Switzerland shows that as long as heaps are not near a water course and in wet areas are covered, there is unlikely to be a problem with liquid run-off from well-made compost heaps.[12] When heaps are made regularly around the farmstead, on specially designated areas, drainage can be built in, to direct the run-off to storage tanks either for use as liquid manure, or for incorporation into new compost heaps as an activator.

Municipal Composting

In industrial societies, where most of the population live in towns and cities, it is essential to recycle the huge amounts of organic waste that we produce in order to help maintain soil humus levels. The production of organic waste in industrial societies today is prodigious. In the USA alone, 270 million tonnes of municipal solid waste (MSW) and 9 million tonnes of dry sewage sludge are generated every year![13] Some 50–60 per cent is organic material, which means up to 162 million tonnes of compostable material per annum! To achieve the aim of nearly 100 per cent organic recycling, composting plants are essential. Fortunately there has been an encouraging growth in both municipal and privately run composting plants in Europe and the USA in recent years. In 2003–4 in the UK, 1,972,000 tonnes of municipal waste was composted, but this is still only a fraction of the 18 million tonnes produced annually, so there is a long way still to go.[14]

It is worth mentioning here that there are heavy metals in sewage sludge, which is why its use is banned on organic farms in the UK (which will be discussed later). At the moment in the UK, most municipal composting comprises garden waste, with as yet only a small proportion made up from waste food, and none using sewage sludge as an added activator. The resultant product is valuable for increasing soil organic matter thus improving soil structure, supplying slow-release nitrogen, improving and stimulating soil microbiology—as does all good compost; but without the huge amounts of nitrogen-rich material in both waste food and manure (both human and animal), we are missing out on a huge natural resource. Of the UK's 1,972,000 tonnes of composted material mentioned above, approximately 18.5 per cent was used as mulch, over 21 per cent was used as soil conditioner, 20 per cent was used to cover over and reclaim land that had been used for landfill, and 40.5 per cent was used to improve agricultural land.[15] This sort of product will always be valuable, but we need to do more.

To produce compost of a high grade, the same principles as apply to the manufacture of good farm compost must be adopted. It must include all properly treated household, catering, factory and institutional waste food, and uncontaminated sewage sludge wherever possible, with regular carbon to nitrogen ratios of between 15 to 1 and 20 to 1, with a mix of garden waste, leaves, sawdust, woodchips, horticultural coarse waste, soiled waste paper etc, with regular carbon to nitrogen ratios of between 40 to 1 and even up to 700 to 1. If this were done, massive quantities of valuable compost would be available for agricultural and horticultural use. Several facilities in New York State are using horse, dairy and poultry manures mixed with sawdust or woodchips to lower the moisture content and increase the carbon content, thus producing valuable compost. They are also collecting and composting two other types of organic waste: leaves and park and garden refuse, and food waste and food-soiled paper from groceries, restaurants and homes.[16]

Large-Scale Composting Methods [17]

There are four types of large-scale composting: the windrow, the aerated static pile, the in-vessel, agitated bed and silo type, and the fully contained rotating system.

The *windrow system* is the easiest and cheapest way of composting on a large scale, and consequently the most common. It is similar to farm compost heaps. Rows of compost are built, typically 2–3 metres (6–10 ft) high and 5–6 metres (16–20 ft) wide. They are regularly turned mechanically both to incorporate air and to keep the temperature from rising above 60°C which would start to kill the valuable bacteria. The drawback of this system is the area required, as compared to other systems. Although there are considerable capital and running costs to this form of site, it remains the most popular form of operation in the UK, which suggests it is the most economic to run.

In the *aerated static pile*, the heaps are similar, but they have a perforated pipe running underneath, through which air is pumped. The system is controlled by a thermostat, which switches it on when the temperature exceeds the optimum. In this system the compost is not turned, and it tends to be more appropriate for the composting of sludges from the food and brewing industries mixed with sawdust or woodchips.

In-vessel systems are production line systems, with the material arriving at one end and the finished product coming out at the other. As in the aerated static pile, in-vessel systems are more appropriate for sludges and waste food. The most common is the silo type, which is a large silo or vessel. The air is pumped up through it with the waste continuously added at the top and compost extracted at the bottom. The agitated bed system is similar, in that it is also contained. The material in this case is passed through by the action of paddles from one end to the other, with air passing up through the mix as before. The paddles not only move the contents forwards but continuously mix and agitate them.

The *fully contained rotating system* is like a huge, long perforated revolving drum, which both aerates and sieves the contents as it moves through. At the end, various grades of compost are produced.

All the last three systems, although more expensive, have the advantage of keeping the smells and diseases associated with sludge or waste food down to a minimum. The plants can be housed reasonably near residential areas. Moreover, in all four types, optimal composting conditions can be monitored and maintained. This means:

Oxygen 10 per cent
Moisture 40–60 per cent
Carbon to nitrogen ratio 30:1
Temperature 32°–60°C (in-vessel plants are processed at higher temperatures to eliminate diseases)

As in most countries, most municipal composting in the UK is done in the open in windrows. They accounted for over 80 per cent of all large-scale projects in 1999, with in-vessel technology at 4.5 per cent and agitated bed and fully contained systems combined at 4 per cent.[18] The figures for 2003–4 are similar, the major change being to the percentage of in-vessel plants, now at 12.9 per cent.

Siting and Collection

There have to be planning requirements and restrictions for new composting sites, in order to protect local residents from noise, smells and undue traffic, but with thorough and honest information and consultation with local people this can be achieved. With the latest organic odour-control filters installed in closed units, smells can be eliminated.[19] The public are aware of the problems with incinerators and landfill sites, so if they are informed about this valuable and essential alternative strategy, then they are more likely to support it.

The planning authorities also have to be more imaginative and adaptive, and this requires a lead from central government which is lacking at the moment. Obtaining planning permission and licences takes too long, and licensing conditions for contractors are too rigid.[20] When a culture is created where recycling and sustainable practices are given priority at both central and local government level, it will be much easier to get both local authorities and the general public on side.

In a lot of instances, it will not be necessary to find new sites. Existing landfill sites are in most cases ideal for such projects. Instead of continuing to bury most of the waste, landfill sites could be used for sorting out recyclable materials. At the same time they could be composting waste food (in-vessel), garden waste, collected leaves and other organic waste in a separate part of the site, with no more disruption and smell than was created when used as landfill.

Farms are also ideal sites for composting enterprises. They have the advantage of not being so restricted by planning consent; indeed, in 1999,

85 per cent of the on-farm sites in the UK were exempt from licensing, and only 50 per cent needed planning permission.[21] In 2003–4, those needing planning permission had increased to 78 per cent. However, this has not restricted the growth of on-farm enterprises. Those requiring licences are the larger sites that process over 400 tonnes at any one time. The advantage of on-farm sites is that they offer the farmer opportunities for diversification. For most farms the size is no problem, existing machinery may very well be suitable, and some or all of the compost can be used on site. Although the farmer may run the sites (and about 20 per cent are sole operators) 80 per cent are run by waste collection authorities, waste management companies or local authorities, or in some cases agricultural companies.[22] For more rural areas, or where towns are in rural surroundings, or on the edges of larger towns and cities abutting countryside, on-farm composting seems one of the most obvious and elegant of solutions – a local solution to a local problem.

Disease Control

Epidemics of both swine fever and foot and mouth disease in the UK resulted in the government introducing The Animal By-Products (Amendment) Order 2001, which put restrictions on the composting of both catering and household organic wastes. Fortunately this restriction has been superseded by the EU's draft Regulation on Animal By-Products, which allows the use of properly composted mixed waste (in-vessel) on all land except pastures. In other words, compost made from waste food can now be used for the growing of horticultural and arable crops, but not where animals graze. If there are problems in the future, it could always be treated in a similar way to some sewage-sludge composting, in an in-vessel plant where the temperature could be maintained at a high enough temperature for as long as necessary to ensure the destruction of most pathogens.[23] Although in-vessel composting has risen to 12.9 per cent in 2003–4, it is still not enough to deal with the amounts of waste food that needs safely processing. It is understandable that after the recent disastrous epidemics, legislation such as this was introduced, but waste food, with its high nitrogen content and carbon to nitrogen ratio of 15:1, is an invaluable activator, available in huge quantities, that should be used to break down coarser waste to provide top-quality compost for food production. In the UK, in-vessel systems will have to increase dramatically over the next few years to process waste food and other animal and waste vegetable matter, both because of the need to keep disease at bay under the Animal By-Products Regulations, and to satisfy the next stage of the EU's landfill regulations, which require that all catering waste and other waste food containing animal by-products be banned from being buried in landfill by December 2005. At present the number of these facilities is way behind schedule.

Treatment facilities to process catering food waste to safe standards are increasing in the UK, but not at a rate that will process all the waste by the end of 2005. This is clearly a matter that needs urgent attention.

Collection and Sources
Sophisticated machinery is now available which can extract non-organic materials such as plastics, glass and metals from mixed waste, and there are plants that can compost the remaining organic material. However, such systems are expensive to both build and run, and there is evidence that the finished product becomes contaminated by contact with heavy metals and other 'nasties', as well as by small non-organic contaminants. Studies have shown the amounts of contaminants to be consistently less in compost made from waste separated at source than in mixed waste.

The obvious answer therefore, is to encourage householders and businesses to separate their wastes before collection. Experience shows that to encourage local participation, the authorities need to supply obviously marked bins or bags to our doorsteps. Ideally same-day collection is best, but if not, there need to be clear instructions, and consistency, in the days when the different collections will take place. Fortunately this is increasingly happening in the UK. In 1999 it was only 10 per cent, but in 2003–4 it was nearly 29 per cent, and growing fast. The sooner it is 100 per cent, the better.

Austria, the Netherlands, and Denmark are already recycling 50 per cent of their household organic waste, and they would do more if they did not have so many electrical plants burning the waste. All three countries have separate waste bins at the door, supplied by the local authorities. Similar schemes are being introduced in the UK; in the USA it has been found that in areas where recycling is most successful, it is because all the different types of rubbish are collected on the same day, even if by different companies. 'Keep it simple, make it easy' is their catchword.

EU Land Directive

The EU Land Directive came into force on 16 July 1999.[24] Its purpose is to reduce the amount of landfill in all EU countries to 75 per cent of 1995 levels by 2010, to 50 per cent of 1995 levels by 2013, and to 35 per cent of 1995 levels by 2020.

These targets are pretty unambitious, but they are better than nothing. Let us hope it means increasing amounts of composting, rather than burning the materials and creating even more carbon dioxide and pollution. The UK's own Waste Strategy 2000 aims to recycle or compost at least 30 per cent of household waste by 2010 and at least 33 per cent by 2015.

There is a lot of work to be done to meet these targets. To encourage members to increase the range of municipal and other large-scale composting schemes, the EU has produced the EC Working Document—Biological Treatment of Biowaste.[25] It aims to provide a framework to promote the production of good quality composts that can be marketed, whilst at the same time placing constraints on lower-grade materials.

Member states are called on to set up separate collection schemes to collect biowaste separately from other waste. Biowaste includes food waste (from households, restaurants, canteens, schools and public buildings), other organic waste (from markets, shops, small businesses, and commercial and industrial sources), and green and wood waste (from private and public parks and gardens).

The Composting Association

The Composting Association was formed in 1995. Its main purpose is to represent the composting industry, but it also promotes composting in general, encourages research and promotes best practice, and acts as a central resource for disseminating information. It also aims to provide a united voice for composting in the UK, speaking to central and local government about the benefits. In 2004 the Composting Association had over 700 members from all sectors of the UK waste management industry, including compost producers, local authorities, consultants, trade suppliers, compost users, academics, individuals and students.

The Composting Association publishes annual surveys to quantify the state of the composting industry, and composting in general, in the UK. The results of the 2003-4 survey show there were 325 operators running 325 sites, processing approximately 1,970,000 tonnes of material, as compared to the 1999 figures of just 90 operators running 197 sites, processing approximately 833,044 tonnes of material—an average growth of around 20 per cent in recent years. Of the 1,525,000 tonnes of municipal waste composted, 54.2 per cent was garden waste taken by householders to municipal waste sites, and 18.3 per cent was household waste from kerbside collections. This is a dramatic change from the 1999 survey, where only 7.5 per cent was kerbside collections and 72 per cent had to be taken to sites by householders. This shows the growth in recycling schemes as part of the normal dustbin collections. However, we still need to urgently increase composting of all organic waste, and it is kerbside collections that still need most expansion at the present time. The vast majority of household waste, consisting of garden waste, is still brought by householders themselves to collection sites.

The growth of this industry has been steady over the last decade, with a low in 1990 of only 4 sites, 32 sites in 1995, 100 sites in 1999 and 325 in 2003-4. This is a growth rate of just over 20 per cent per annum at the moment.[26] This is healthy, but it needs not only to be sustained, but to be speeded up.

There is an equivalent organization in the USA called the United States Compost Council (USCC), whose mission is very similar to the Compost Association, in that they encourage and guide research, promote best practice, establish standards, run educational programmes for professionals as well as educating the general public, and encourage improvements in the quality of the compost and the development of markets for the final product.

Uses of Municipal Compost

As I have said, the uses to which municipal compost are put tend to be rather limited at the moment, mainly as a mulch and soil-conditioner. However, experiments show that it is valuable as a constituent of potting compost when mixed with other products such as ground coir fibre. There will always be a problem with this sort of compost: its lime content means it can only be used for plants that are lime-tolerant. Nonetheless, municipal compost could help to eliminate the use of peat, the mining of which is ecologically damaging and unsustainable. In the UK in 1999, 36 per cent of all municipal compost was used as mulch, 35.5 per cent as soil conditioner and 14.3 per cent for land reclamation, mostly to cover over landfill sites, and occasionally disused gravel pits, slagheaps or old industrial sites. That left only 9.3 per cent for the production of a growing medium for horticultural use, and 2 per cent for the creation of topsoil and other uses.[27] The dramatic change that has occurred is the increased use of municipal compost by farms, largely due to a huge increase in on-farm composting sites since 1999, an increase from 65 sites in 1999 to 246 in 2003-4. At 47.9 per cent of the total compost produced this amounts to 570,000 tonnes now being used for agriculture and horticultural use, whereas in 1999 it was only 8 per cent, or 66,401 tonnes. This trend is very encouraging.

However, as production increases and separated kerbside collection is better established, much more high-grade compost will be made, with the addition of high-nitrogen products such as waste food, which could be used for crops. One innovative use is as a decontaminant. The highly active micro-organisms in a mature compost can be used to both sequester and break down contaminants in soil and water. In soils contaminated with lead and other heavy metals, abundant applications of compost were found to bind the lead, making it inactive. As an experiment, lead-contaminated soil was fed to one group of rats and the same soil mixed with compost to another. The rats fed soil alone showed signs of lead poisoning whilst those that had added compost in their meals showed none.

At the Seymour Johnson Air Force Base in North Carolina there are regular oil and aviation fuel spills. The contaminated soil is removed and mixed with compost and turkey manure and placed in heaps. It is turned mechanically to keep it aerated, and the fungi in the compost produce a substance that breaks down petroleum hydrocarbons, enabling the bacteria in the compost to metabolize them. Once treated, the soil is put back. Whether this is completely effective is not known, but this use of compost has great possibilities for dealing with such problems.

Stormwater which has run off roads, roofs, car-parks, lawns and gardens contains a variety of pollutants such as heavy metals, oil and grease, pesticides and fertilizers. To comply with the US Environmental Agency's National Pollutant Discharge Elimination System Regulations, some US municipalities and industries are starting to use composting technology. Tanks are built into the stormwater pipeline, with a layer of special compost, largely leaves, through which the water has to pass. These units have been found to remove 90 per cent of all solids, 85 per cent of oil and grease, and between 82 and 98 per cent of heavy metals from the stormwater. Moreover, they are able to treat up to 0.23 cubic metres (8 cubic feet) per second!

The most ingenious of all, however, is an indoor composting facility in Rockland County, New York, which had problems obtaining planning permission because of potential smells. As a result the designers came up with an odour-control filter which first passes the expelled air through an ammonia scrubbing unit and then forces it through an enclosure filled with compost and other organic materials. The compost binds the odours which are then degraded by the micro-organisms.

Many more uses for compost as a pollutant control agent will undoubtedly be discovered in the future. Mature compost houses vast populations of valuable micro-organisms as well as a unique ability to bind with a whole host of different chemicals and compounds.[28]

Sheet Composting, Green Manuring and Zero-Tillage

One of the criticisms of the sole use of compost to feed the soil is that the fermentation process takes place away from the soil itself, and while this encourages mycorrhizae fungi and micro-organisms when added to the soil, their growth is even more prolific when they have to deal with unrotted or partially rotted organic material on and in the top few centimetres of soil. An extra benefit is that the breakdown of unrotted organic matter in the soil has been shown to increase the inherent disease resistance of the soil.[29]

The counter-argument is that if high cellulose material is incorporated into the soil, the micro-organisms are so busy using the available nitrogen to break down the cellulose that it results in a serious nitrogen shortage for the growing crop. Moreover, in a properly made compost heap, the temperatures attained are high enough to kill off weed seeds and many kinds of plant diseases during the fermentation, and the whole process of breakdown is speeded up. In reality, it is not an either/or argument. Organic farmers use composting as well as sheet composting and green manuring in feeding their soils.

Sheet Composting

Sheet-composting is composting in situ, in other words in or on the soil itself. Fresh farmyard manure can be spread on the land, and a green manure (a crop that is grown solely for the purpose of feeding the soil and stopping the loss of nutrients) is sown. When it has grown it is lightly ploughed or harrowed in. Alternatively a green-manure crop may first be grown and the manure spread on the grown crop before cultivating the soil.

The seven-year rotation we employed on our farm included four years of grass leys, followed by wheat and other grain crops, followed by field beans and finally root and fodder crops for the animals, including potatoes, cattle kale, fodder beet, and swedes. Root and fodder crops require a lot of feeding, so when the previous crop of field beans had been harvested, some beans and a little grazing rye were sown to add to those beans spilt during harvesting. These were then disc-harrowed into the soil. By the spring they were well grown and the beans were providing some nitrogen in their root-nodules. Compost was then spread on the crop and the whole lot—including the stubble from the previous bean crop—was ploughed in ready for the root crops to follow.

Green Manuring

Without the addition of the manure or compost, the bean crop would have been called a green manure crop. In reality, green manuring is a form of composting which is used to supplement the application of compost and manures. They do not usually result in significant amounts of humus for the soil, but they can increase nutrients and feed and encourage soil life. They have the added value of covering and protecting the soil surface and cutting down the leaching of nutrients, especially during the winter months.

Zero-tillage

The practice of zero-tillage has grown at almost exponential rates in the last few years in countries like Argentina, Paraguay and southern Brazil, and is increasingly being adopted in various forms in Europe and the USA. However it is not a new idea. Masanobu Fukuoka, a Japanese microbiologist, who took over his father's farm after the Second World War, published his now famous book, *One Straw Revolution*, in 1975. In it he documents the agricultural method he had developed over thirty years. This involved a 'do nothing' approach, without the use of ploughing, tilling, chemical fertilizers, pesticides, weeding, pruning, machinery or, for that matter, compost! He produced high-quality fruit, vegetables and grains, with yields equal to or greater than his neighbours. He grew two seasonal crops: rice in the summer, and barley and rye in the winter, using just the straw of the preceding crop, a cover of clover and a sprinkling of poultry manure for fertilizer. Instead of sowing the rice seeds and transplanting the seedlings, he broadcast clay pellets containing the seeds onto the unploughed land, covering them lightly with the straw from the preceding crop. The pellets helped retain moisture until the young plants were established, as well as discouraging birds from eating the seeds.

The zero-tillage techniques now being used increasingly in South

America, in particular, are not as radical as Fukuoka's, as they involve the occasional use of selective weedkillers and fertilizers. However, there are many farmers using zero-tillage who are either organic or nearly organic.[30] After the harvest, crop residues are left on the field as protection against soil erosion and for the creation of humus. At planting time, the seed is slotted into a groove cut through the crop residues and into the soil surface. This means that the soil surface is always covered, and the soil is never inverted.

Green-manure crops are used as integral parts of crop rotations, for weed suppression, winter soil cover and the improvement of nitrogen where legumes are used. One of our neighbours in the next village, who grows a lot of cereals by local standards, adopted no-ploughing techniques a few years ago. He harrows the stubble after the harvest with spring-tine harrows, adding partially composted bedding straw and cow manure to the land. He then harrows the land at least once more before drilling the next crop of grain. He told me that after three years, when the humus levels have built up, yields improved and he made savings in energy because of the ease with which the soil could be prepared for the next crop. Ploughing and the subsequent work in breaking the soil down to prepare a seed bed, requires large amounts of fuel and effort. Although he still uses fertilizers and sprays, this technique is an improvement on more traditional methods.

In many parts of the world, combine harvesters now have equipment on the back which chop and spread the straw at harvest before it falls to the ground. It is then left on the field and the next crop sown through it.

Sewage

Once properly composted, any manure, however revolting when fresh, becomes a sweet smelling, dark, crumbly humus. However, one of the standards by which we measure our civilization, is the sanitary way we dispose of our sewage, which usually means that we are wasting huge amounts of invaluable organic fertilizer and at the same time polluting our seas, rivers and atmosphere.

The problem is that sewage is added to rainwater and industrial waste water, and this makes the whole process of treatment and recycling much more complicated—although new building regulations in the UK are tending to separate rainwater from sewage. There are three problems: an excessive water content, the pollution of the sewage by contaminants from both industry and rainwater run-off from roads (which includes heavy metals

Maintaining Soil Fertility

such as lead, copper, cadmium and chromium, as well as pollutants such as chlorinated hydrocarbons), and the way sewage is treated, which wastes vast amounts of valuable nutrients.

An Invaluable Asset

Sewage is a huge wasted asset. In the conventional sewage treatment process, vast volumes of carbon dioxide, ammonia, gaseous nitrogen and methane are released into the atmosphere, contributing to global warming. Once disease pathogens and pollutants are eliminated, sewage should be recycled back onto the land. This is one of the biggest nuts we have to crack.

F. H. King, in his book *Farmers for Forty Centuries*, cites research from two scientists, Carpenter and Hall, who estimated the manurial value of the sewage produced by 1 million people per annum. An average adult produces 1.14 kg (2.5 lb) of excreta (liquids and solids) per day, which makes 413,884 tonnes per million people per annum. According to Carpenter's figures, this would provide:

2,631 tonnes of nitrogen
829 tonnes of potassium
352 tonnes of phosphorus

Hall calculated the content as:

3,605 tonnes of nitrogen
1,394 tonnes of potassium
892 tonnes of phosphorus [31]

With a population of about 58 million, that means that each year the UK is losing:

180,844 million tonnes of nitrogen
64,496 million tonnes of potassium
36,076 million tonnes of phosphorus

These figures are obviously approximate, but they show just what the recycling of human sewage could do for the production of humus and the feeding and maintenance of soil fertility. At the same time, they make clear the scale of waste and pollution caused by the present way we deal with our sewage.

Sludge

The most common way of treating sewage results in a sludge that is a mere 6 per cent of the original material, and with a meagre nutrient content of: 2–3.2 kl (440–700 gal) per tonne of nitrogen, 0.8–3.0 kl (175–660 gal) per tonne of phosphate, and 0.1–1.8 kl (22–400 gal) per tonne of potash.[32] In other words, the nutrients in the sludge amount to about 2 to 3 per cent of the original fresh sewage. The rest are wasted to the atmosphere.

In many countries a proportion of the sludge is then buried in landfill sites or burnt, the latter being both very expensive and polluting. However, some countries are starting to use sludge as an activator to help compost coarser materials such as leaves, garden waste, waste paper, straw and sawdust. Sawdust is very valuable because it helps to absorb the large amounts of water in sludge, typically 70–80 per cent. Sewage sludge composts have to be carefully maintained at high temperatures (55°C) to destroy disease pathogens.[33] Other sludges, such as brewery sludge, paper mill sludge and food processing sludges, can be composted in the same way. All have a high water content, but are nonetheless a valuable addition in the process of humus creation.

Heavy Metals

Tests at Rothamsted Experimental Station have shown that the regular use of sewage sludge over a twenty-year period—admittedly well above the levels now allowed—reduced soil organisms to 50 per cent or even 30 per cent of normal, owing to the presence of heavy metals. Blue-green algae were also reduced substantially, leading to poor nitrogen fixation of the soil. Heavy metal contamination in soils will last for thousands of years and pollute the food grown on it, so this is a problem that has to be dealt with.[34] This is why at the present time, the Soil Association's rules do not allow the use of sludge for organic production, although a high proportion of London's is used today on conventional farms around the capital.

The heavy metals in sewage sludge come mostly from water from industrial processes, from lead on roofs, and from the now-banned lead in petrol, which found its way into the drainage systems. There is no way of eliminating heavy metals, so the only way to solve the problem is to try to remove or reduce the pollution in the first place. In many countries, industry is subject to increasingly tighter laws requiring the removal of certain contaminants before releasing the water into municipal sewers. One such is the US

Environmental Agency's National Pollutant Discharge Elimination System Regulations. These include the removal of copper, lead, cadmium, chromium and chlorinated hydrocarbons. As we have seen, some US municipalities and industries are starting to use compost technology to solve the problem. This approach is starting to have an effect in reducing sewage contamination, with the result that many more municipalities in the US are looking more favourably at the composting of sewage sludge and its subsequent use on land.

Forestry

In Nelson, New Zealand, the municipality pumps its sewage sludge onto a forested island, which both deals with the sewage and helps to produce useful timber. Pollutants still need to be eliminated, but this sort of imaginative use of sewage needs to be considered more and more. Forestry use also has the added advantage of overcoming the common reluctance to use sewage sludge on crops.

Methane Production

We have to accept that the existing methods of sewage treatment are going to continue to be the norm. Under this system, the sewage is fed into settling tanks which are aerated through agitation and at a later stage the liquid effluent is sprinkled onto gravel beds which contain high populations of bacteria as well as oxygen. The carbon is given off as methane and carbon dioxide and the nitrogen as nitrous oxide and ammonia, so much of the manurial value is wasted to the atmosphere. With further cleaning, the resultant water is clean enough to be released back into the waterways, or sometimes even into the drinking water system.

However, sewage plant sites could be adapted to use more ecological methods, for instance methane production. Indeed, there were sewage plants in Victorian England that used this technology, and some sewage plants are reviving the practice.

Under this system, instead of aerating the liquid effluent, it is contained in digesters in anaerobic conditions, thus retaining the nitrogen content, whilst the carbon is broken down by bacteria into methane. The carbon and hydrogen from the decayed cellulose combine into a gas, which can be used as a fuel either for direct burning or to produce electricity using direct fuel-cell technology or even, in the near future, synthetic diesel directly from the methane.

After the digestion process is completed, a rich nitrogen slurry is left for use as a fertilizer or as a more powerful activator than sewage sludge in the production of compost. The advantage of this kind of technology is that none of the nitrogen is lost, and that the water vapour and carbon dioxide given off when the methane is burnt only replaces the carbon dioxide absorbed in the production of the food eaten in the first place. This technology is therefore a vast improvement on conventional processes, saving on fossil fuels and therefore on the production of more carbon dioxide.

In many Third World countries this kind of technology is already increasingly used on a local scale. Jules Pretty, in his book *Agri-Culture*, writes about an ecological demonstration village in northern China, in Yanqing County, where instead of monocropping maize as they had previously, they have now diversified into growing vegetables and keeping pigs and poultry, and are saving on precious fuel. Each household has a biogas digester, which produces methane gas for heating, cooking and lighting. All their green waste and waste food is fed to the animals, and their waste goes into the digester. The remaining solids are used to fertilize their plots of land. Already there are 8.5 million households with digesters, and the government wants to introduce this system to 150 counties across China, with a target of 1 million new digesters per year. In many country communities in China, wood and the woody stems of crop residues were traditionally used for fuel; now each digester saves the equivalent of 1.5 tonnes of wood per year.[35]

In Nepal there are 60,000 small-scale biogas plants, and in many other countries, including India, Malaysia, Morocco and Thailand, they are playing an increasing role. In New Delhi 4,000 public toilets have been built, producing biogas and fertilizer.[36]

Methane can be used to power fuel cells to make electricity. In Washington State, USA, the King County government in partnership with Danbury Inc, a Connecticut-based FuelCell company, has a 1-megawatt direct fuel cell power plant running off sewage-produced methane, making electricity for the local grid. They are using this project as a showcase, and are looking to convert all their sewage plants to produce biogas for local electricity production. FuelCell Energy Inc. also have a more recent project involving a 250-kilowatt plant at the Terminal Island Treatment Plant for Los Angeles.

At the University of California, Riverside, successful experiments have been carried out converting sewage sludge and grass clippings into synthetic diesel fuel from the methane that is produced, sponsored by the local public utilities and the Eastern Municipal Water District. The slurry mixture is pumped into a steam generator that heats it to about 700°C at 30 atmospheres pressure. The super-heated waste is then mixed with hydrogen gas inside a reactor, which produces methane. The methane and the superheated steam are then fed into a second reactor, in which they react to produce hydrogen, carbon monoxide and carbon dioxide. Half the hydrogen is recycled back to the first stage, making it self-sustaining. The gases are then passed to a liquid fuel synthesizer, designed to produce sulphur-free synthetic diesel fuel, electricity and recycled clean water. Molten salt which is heated in the process carries the heat back to the water steam generator and the second reactor, making the plant almost thermally self-sufficient, and the electricity is used to help run the plant. Thus carbonaceous waste and waste water can be converted into fuels to process heat and decontaminate water, in what is expected to be a series of largely self-sustaining processes.[37]

New Housing Estates

It is where new housing estates are built that the most progressive systems of sewage treatment have a chance of being implemented. The best water-closet system would comprise low-flush toilets, with a separate piped system separate from other household waste water, flushed by saved rainwater and piped to a local small-scale sewage plant to be either mixed with high-cellulose waste and carefully composted, treated in a reed-bed system (see below), or digested to make bio-gas for the benefit of the local community. The product would not be polluted with heavy metals etc., and would be much easier to deal with, because of the much lower level of water. Indeed, in the UK there is increasing separation of run-off water and sewage.

Reed Beds

Reed-bed technology is a more ecological way of treating sewage, and is becoming more popular with local authorities. In the UK Prince Charles has helped to make reed-bed systems more acceptable by installing an extensive system at his Highgrove estate. The system consists of a lined and sealed bed of gravel, sand or soil in which reeds are grown, usually *Phragmites australis*, the common reed that grows in the fens of England and is used for thatching. In most systems the sewage effluent is piped under the gravel, although it can be spread on top.

The reeds have two functions. First, the very extensive root system creates channels for the water to pass through. Secondly, the roots introduce oxygen into the body of gravel or soil and provide the ideal environment for aerobic bacteria, which break down the sewage. This is a highly effective method for ensuring that the water passing out of the system is of very high quality. However, in terms of recycling the nutrients it is not perfect. In many ways reed-bed systems are just a greatly improved, organic version of conventional sewage treatments, with the same problem of a substantial loss of nutrients to the atmosphere. On the other hand, in the spring and summer 15 per cent of the nutrients are taken up by the plants themselves, which can be harvested on a regular basis for composting, thus recycling some of them. By enclosing the reed-bed under glass or plastic, the growing season could be extended, increasing the conversion of nutrients into plant material for composting.

Combinations

An even more productive approach would be a combination of methods such as settling tanks, a biodigester, reeds and compost. Fresh sewage would be passed into settling tanks, allowing much of the solid matter to settle. A large proportion of the liquid would then be fed into a reed-bed system, and the more solid material into a biodigester. The methane produced would be used to run the plant, including heating the digester and improving its efficiency, and the nitrogen-rich slurry from the digester would be used, with the cut reeds and other high-cellulose waste, to make compost.

New Imaginative Thinking

With flexibility and imaginative thinking, it is perfectly possible to recycle most of our organic waste, thereby providing invaluable humus and considerably reducing pollution of the atmosphere, rivers and seas. A determination to seek solutions helps to overcome obstacles. For instance, if composted waste food cannot be used on pastures because of concerns about spreading animal diseases, then we could use it on arable fields. If there are objections to using composted or treated sewage on crops grown directly for human consumption, then it can be composted safely, or used on pastures, on crops grown for animals, or in forestry.

Chapter 7

Control versus Co-operation

I am the one whose praise echoes on high.
I adorn all the earth.
I am the breeze that nurtures all things green.
I encourage blossoms to flourish with ripening fruits.
I am led by the spirit to feed the purest streams.
I am the rain coming from the dew
that causes the grass to laugh with the joy of life.
I am the yearning for good.
—St Hildegard von Bingen (1098–1179)

How Did We Get Where We Are Today?

Was it scientific discovery and technological innovation that led to the kind of modern conventional farming we see today? Was it the exponential increases in population that have occurred around the world over the last 200 years that forced countries to increase their food production by any means? Both these factors certainly contributed, but is it not possible that the causes are much older and more profound than that, and that they stem directly from particular philosophical and scientific paradigms?

I am referring to the way we perceive Nature in Western culture, the way we perceive our place in the scheme of things, what we feel is important to us, and our accepted beliefs and world view. These have had a profound influence in bringing us to where we are today. It is our attitude towards Nature, above all else, that is, I would suggest, one of the most important factors in defining the way we have farmed throughout history in the West.

Which came first? Did the industrialization of the West, with its subsequent massive population movements which cut off the majority of people from the countryside, produce an alienation from Nature, along with the industrialization of farming that created a barrier between the farmer and the soil? Some have argued that this has been the main cause for the increasing feeling of separateness from Nature in modern society. These trends have surely played their part in our changing attitudes, but one also has to take seriously the proposition that it was the changes in perception about our relation to Nature that came first, and it was these that governed our subsequent behaviour. Over the centuries, the way we perceived Nature and our relationship to it has changed dramatically.

Why is it that we perceive Nature as separate? Why have we seen Nature as uncontrolled, anarchic and something to be feared, or at the very least something to be wary of? It is no coincidence that popular programmes on TV often have titles such as 'Angry Nature', and are about tornadoes, eruptions, earthquakes, tsunamis, floods and hurricanes, all of which are beyond our control and imply that Nature is out to get us!

This attitude is also adopted by gardeners. My father was a keen gardener and often said to me, "Gardening is a form of warfare", referring to his constant fight with diseases, weeds, bugs, birds and cats. I protested that in my experience it was about working *with* Nature, but he could not agree, despite the fact that he was a country lad with a great love of Nature, among whose favourite poets were Wordsworth and Clare.

It is because for centuries we have perceived Nature as separate and threatening that we have wanted to dominate it. Indeed, it is largely because Nature is seen as separate that we are fearful of it and wish to control it. Once this philosophical mistake was made, it was inevitable that it became compounded over the centuries. Fortunately it is at last starting to be questioned.

To understand the growth of the organic movement in the last century one has to understand how we got to where we are. I want to try to trace the history of our Western view of Nature, and to compare it with other cultural and historical views, in an attempt to understand how we arrived here and where we should go now. Of course I recognize the answer is not simple, as many different historical forces have come into play over the centuries, but I have come to the conclusion that a common thread runs throughout the history of Western farming, and that is our eccentric attitude to Nature.

Although European culture did not develop in strict isolation, and there were trade and cultural links with both Islamic and Oriental cultures during the Dark and Middle Ages, there were few challenges to the ideas and paradigms that developed during this period of comparative isolation. When

Europeans did eventually start making more contacts again with other cultures, it was as conquerors, traders or missionaries, who were there to 'educate' and impose their ideas and beliefs on others, rather than trying to learn from them. You cannot learn from other cultures if you believe those cultures to be inferior.

> It is important to make clear that the word 'Nature', as used in this book, does not denote Nature as Goddess, Deity or indeed any form which would imply a life of its own, simply because that would once again make it something separate, something 'other', something 'out there'. The capital 'N' is used to denote Totality/Unity—that which contains everything and includes everything—it is inclusive, not exclusive. It is that which we are—our essential Nature, not the source of life, but Life itself.

What is Nature?

Before we proceed further, we need to define what we mean by Nature, not only the different ways we use the word, but the connections between its different uses. A good start would be to look at the dictionary definitions. The word comes from the Latin *natura*, literally meaning birth or the course of things, hence:

a) the essential qualities of a thing
b) the inherent and innate disposition or character of a person (or animal)
c) the general inherent character or disposition of mankind
d) an individual character, disposition
e) the inherent power or force by which the physical and mental activities of man are sustained
f) the creative and regulative physical power which is conceived of as operating in the physical world and as the immediate cause of all its phenomena (late Middle English)
g) contrasted with art
h) the material world, or its objects or phenomena, the features and products of the earth itself, as contrasted with those of human civilization

Definitions a, b, c and d all have the same general connotation, namely essence, essential nature, our own nature. In many cultures, this human 'essence', which has many names, was recognized as the essence of all things. Separateness and alienation from other beings and from Nature in general was an anathema to such understanding. This nature which is my essential nature is the same nature as that of my fellow human beings, this animal, this plant, this tree, this earth, this sky, this universe—they are not superior or inferior, but equal.

Definitions e and f are fascinating in that the idea of 'essence' is seen as the initiator, active and creative and at the same time life-sustaining. This view of Nature is also consistent throughout many cultures and times. In Vedic culture it is *Prakriti* (or *Prakruti*), often translated as 'Nature', but better understood as 'Great Nature', the principle of creativity, or the nature of all things. The original Latin meaning of *natura* as both 'birth' and 'the course of things' is closest to this idea. Nature in this context is seen as self-regenerating, and self-sustaining—functioning through a process of self-referral. And once again there is the idea of that essential nature shared by all phenomena, including of course, ourselves.

Definition g is very revealing, as it derives from the Age of Enlightenment. The Oxford English Dictionary dates the first known use of the word in this way to 1704. Enlightenment thinkers were increasingly perceiving the reasoning powers and creative activities of humankind as distinct from the natural world, which in turn was increasingly being seen as the product of random events based on intrinsic natural laws. There was also a perceived difference between human civilization and both Nature and barbarism; in other words, the idea of a 'civil' behaviour that defined human nature at its best—not realising that our creative and reasoning powers, as well as our ability to be 'civil', are also an integral part of Nature. 'Contrasted with art' suggests not only an increasing separation from Nature, but also a growing sense of superiority.

Definition h is Nature as most commonly understood nowadays, meaning 'the natural world'. It is almost as recent in its use as the last example, dating from the seventeenth century. This is again highly revealing. Once again there is the suggestion of the separateness, exclusiveness and even the superiority of human 'civilization'—ideas very characteristic of the Enlightenment.[1]

Inferior to Nature

A more recent view is that of feeling inferior to Nature. Kenan Malik, in his book *Man, Beast and Zombie*, suggests that this loss of confidence in humanity occurred after the Holocaust in the Second World War. But ecologists take the view that this change of heart has come about as a result of a growing understanding of the threat that our activities pose to the environment.

In this perception, Nature is seen as untouched and unsullied by humankind, as though we are not meant to be part of the plot. We therefore feel guilty and even apologetic for our presence on Earth. At its worst, some ecologists believe humans are an aberration of Nature, even a cancer, with irreversible flaws that will inevitably lead to our demise.

Feeling inferior to Nature is just as isolating and unreal as feeling superior; in fact it is the reverse of the same coin. It suggests an inability to choose alternative behaviour, to change our ways, and implies that we will eventually annihilate not only ourselves, but many of the other species as well. It seems to me that feelings of superiority or inferiority are both based on the perception of ourselves as separate. Karl Marx expressed the position in his *Economic and Philosophic Manuscripts*: "That Man's physical and spiritual life is linked to nature means simply that nature is linked to itself, for Man is part of nature."[2]

In seeking to know our own nature, we often objectify 'nature', seeking it outside ourselves in the 'natural world'. That is why there is a great tradition of seeking solace in natural surroundings and temporarily leaving our man-made environments to recharge our batteries by enjoying a holiday in the countryside, or travelling abroad to wilder areas. Is it not possible that the experience we have of the beauty of Nature is really one of recognition?

What we recognize in Nature 'out there' is the essential 'nature' of all phenomenal existence, which is also our own essential 'nature', or Self—that Nature which is both the essence of existence, and at the same time its physical end product.

The 'Ancients'

Before the advent of agriculture and cities, around 8,000 to 10,000 years ago, humans were tribal hunter-gatherers. By studying the less than one per cent of the world's population who still live this way, we can at least gain some insight into how these people perceived the world and Nature. All existing tribal peoples have a common perception of themselves in relation to the rest of Nature. The Saami peoples of the Norwegian Arctic have a word for Nature, *lotwantua*, which means 'everything is included'.[3] As Thom Hartmann said in his book *The Last Hours of Ancient Sunlight*: "Instead of the story that we're 'separate from Creation and born to dominate it,' these older cultures hold a different notion of the place where humans stand in the order of Creation."[4] Tribal peoples know they are dependent on their environment, and view all of Creation as sacred. There is no belief of separateness from Nature, no feeling of superiority or inferiority to the natural world. They have a great respect for animals; indeed in their eyes each plant and animal has its own intelligence and spirit, and they see the whole of creation, including what we perceive as inanimate, as alive.[5]

This ancient lifestyle is sustainable. These people do not destroy their environment, because they know they are entirely dependent upon it. Their cultures are co-operative, both with one another within the tribe, and with their environment. For example, the Cree have traditionally appointed stewards to manage beaver stocks in eastern Canada. They oversee the rules of hunting beaver with their knowledge of past hunting patterns and the abundance of beaver numbers. Without Cree control, the natural chain of events would see huge swings in the beaver populations, because these populations would reach a state where they start to denude their food stocks of willow and aspen, causing a crash in population for many years until it recovers again. Managed hunting ensures that the population is maintained at a consistent level.

The Pueblo Indians have an understanding of the importance of our relationship with Nature in maintaining life. They believe that when people lose touch with Nature, their hearts harden and they inevitably lose respect for animals, and ultimately for mankind.

In the Judaic story of the Garden of Eden, we have an idyllic vision of Nature and the true perfection of a hunter-gatherer lifestyle before the fall and the invention of agriculture, and it is often this vision that we refer back to, particularly now that many of us are even more removed by living in cities. Thom Hartmann described his contacts with people of what he called Older Cultures:

> "I've become convinced that our sense of spiritual disconnection started with our Younger Cultures' disconnection from nature. (One metaphor for this was the expulsion of Adam and Eve from the Garden of Eden.) When we decided to separate humanity from all the rest of creation, we created a schism that was deep and profound. When we decided that the world was here for us, separate from us, and it was our holy duty to control and dominate it, we lost touch with the very power and spirit which gave birth to us."[6]

However, we now refer back to a reinterpretation of the idyll in the form of an agricultural past, rather than the lifestyle of the hunter-gatherer. The Israelites, on the other hand, had difficulties with agriculture and the city-states that resulted from a settled way of life. They were travellers and hunter-gatherers, and tended to be very sceptical of Egypt and Mesopotamia, for example. Judaism was establishing itself in opposition to the cults in the area which worshipped the land itself.[7]

The Classical European Period

The Ancient Greeks

It is difficult to place Homer, the traditional epic poet of Greece, accurately in time—estimates vary, but for the sake of argument we can place his birth at around 1100 BC. In his time, Nature was still not quite seen as separate, although there were already indications of a change of view. His description of the forces of Nature are personified as the wills of gods, sometimes malevolent, sometime benevolent. Nature was described in terms of human passions—the whole of the natural world was understood in terms of ourselves. In his poems we find an invocation of the agricultural life as something natural for us, and the feeling of owning the landscape and Nature, because of the ownership of our plots of land. At this time this was the main change in perception between the hunter-gatherer lifestyle and an agrarian society.

Aristotle

This view was developed by Aristotle (384–322 BC). Of all of the Greek philosophers, he was the one who had the most effect on subsequent European thought and beliefs. He was the first person that we know of in the Western tradition who took Nature as an object of impartial study. He studied the way in which things work, and particularly the way in which life emerges and grows out of itself. His view was not that of the contemplative poet who sees Nature as a spiritual source; on the contrary he saw it as something given to us to understand. He still had a sense of us as part of Nature, but at its high pinnacle. He saw our intellect as something unique, although not apart from the natural order. He believed that both animals and plants had something like a soul, by which he meant the principle of organization of the organism.

To Aristotle, the universe was harmonious and ordered, full of purpose and desire. A consequence of this belief was the idea that all objects have a purpose and even a desire to fulfil their purpose in developing to their full potential, as in the example of an acorn. The acorn, as Aristotle saw it, is in essence an oak tree. It becomes an oak tree because that is its destiny; thus it fulfils its purpose and confirms its nature.[8] He saw no distinction between Nature and art for precisely that reason: both have a purpose to develop to their full potential. In art, the 'seed' of an idea grows and finally realizes its potential, as in Nature (an interesting contrast to the Enlightenment's distinction between the two).

Eastern and Western Wisdom

There were many, often contradictory schools of philosophy in ancient Greece, but there is inherent in most of them a strong echo of an older and more ancient wisdom, which perceived the universe as interconnected and unified. This theme runs through much of ancient wisdom both Eastern and Western, and takes for granted that we humans are also part of this unified creation. As Fritjof Capra commented in his book *The Tao of Physics*:

> "The most important characteristic of the Eastern world view—one could also say the essence of it—is the awareness of the mutual interrelation of all things and events, the experience of all phenomena in the world as manifestations of a basic oneness. All things are seen as interdependent and inseparable parts of this cosmic whole; as different manifestations of the same ultimate reality."[9]

This traditional Eastern way of seeing the world, and man's place in it, is exemplified by Chinese landscape watercolours covering a period from roughly the tenth to the seventeenth centuries, such as Hsu Tao-Ning (c970–1051) and much later T'ang Yin (1470–1524). These fantastic landscapes depict the forces of Nature, rocks, mountains and streams and often wind-swept trees—and hidden within there is a hut, temple or dwelling, and the small figures of travellers or fishermen in a boat. In other words, the men and women and their human artefacts are depicted as an integral part of the landscape and of Nature. There is no separation, no alienation.[10]

This perception of the world and our place in it is shared by both hunter-gatherers and early agricultural societies in both early classical European and Eastern cultures. Through the vision of wisdom nothing is separate, all is one. Interestingly these are also the kinds of conclusion that modern quantum physics is reaching. As the modern physicist, David Bohm, has written:

> "One is led to a new notion of unbroken wholeness which denies the classical idea of analyzability of the world into separately and independently existing parts. . . . We have reversed the usual classical notion that the independent 'elementary parts' of the world are the fundamental reality, and that the various systems are merely particular contingent forms and arrangements of these parts. Rather, we say that inseparable quantum interconnectedness of the whole universe is the fundamental reality, and that relatively independently behaving parts are merely particular and contingent forms within this whole." [11]

In Europe, this understanding was lost some time between Homer and Aristotle. This perception of separation from Nature became increasingly enhanced, first in medieval Europe, then the Renaissance and the subsequent Age of Enlightenment. A comparison between the earlier bronze-age Minoan civilization on the island of Crete and the later mainland civilization of classical Greece exemplifies this change. The Minoans saw no distinction between the divine and humanity; our essential nature is divine and therefore approachable within. They saw their own 'nature' and Nature as one in the same. By the classical Greek period, the gods were not only perceived as separate from us, they were relegated to supernatural versions of humans with all their faults writ large. The feeling of separateness from our own essential 'nature' had begun.

With the increasing development of city life, by the time of Virgil (70–19 BC), the idea of the restorative value of Nature as a place to go to change oneself was becoming more prominent. Virgil was greatly influenced by Theocritus and the Alexandrian poets. Alexandria, and the ideas that abounded there, created a turning point in classical thought and ideas. It had a great library, and in this atmosphere of learning the contemplative poets flourished. The second librarian at the library of Alexandria was Callimachus, who was the founder of the school of idyllic poets. Possibly the greatest of the idyllic poets was Theocritus. For him Nature was a great escape from the urban surroundings of this great city. In his idylls he wrote of the natural world as 'our home'—a place where we really belonged and from which we have been dragged, and by going back there we could regain something that had been lost in living in the city.[12]

The Epicurean Tradition

Epicurus (341-270 BC) initiated a whole new way of thinking about Nature. This was the first major break with the past, which led to the creation of a completely materialistic view of Nature and the universe. This view was rejected by Christianity and did not re-emerge until the Age of Enlightenment in the eighteenth century. For Homer, Nature was a force very closely linked to the divine, allowing us in some senses to discover the divine through Nature. The Epicurean tradition, on the other hand, had no place for the gods; it simply said, 'All you have is matter, atoms, and vacuum.' Epicurus and his successors looked on the world and Nature as purely material, without any spiritual dimension at all. The process by which natural forms take shape was seen as involving a kind of randomness. Atoms swerve together and then swerve apart. Human beings were seen as being made when atoms swerve together, and when we die the atoms dissolve back again into the universe. However, humanity's place in the order of things was not seen as special, because just like the animals, we are all made of atoms.[13]

The Romans

The Epicurean tradition filtered over into poetry with the Roman poet Lucretius, who wrote a poem called *De Rerum natura* ('Of the nature of things'), a kind of versification of the Epicurean system. Lucretius warned against religion, because he believed that when one turns one's attention to the supernatural one fills one's head with demons of one's own creation, which he believed was the origin of enmity between people.

Roman poets such as Ovid, Horace and Virgil largely wrote for the Roman court, on subjects such as civility and the virtues of Romanness, but often referred to Nature 'out there', using it in their imagery. Nature was seen as outside the city and city life, and therefore as somehow more natural. For Horace, Nature was somewhere to which one could retreat to be true to one's nature, among friends, away from the artifices of the city.

The Judeo-Christian tradition, which became increasingly dominant in the West, was very suspicious of the Epicurean tradition, with its very mechanistic view of the universe, because it did not leave a place for an immortal soul or a transcendent God. It did not leave a realm above Nature, and consequently did not place mankind at the head of creation. As they saw it, humans also have souls, which do not fit into a universe of natural phenomena, being uniquely in God's special care.[14]

Unsustainable Farming Practices

Throughout this period in Near Eastern and Mediterranean civilizations, as the perception of Nature as separate grew, so did unsustainable farming practices. In Sumeria, a combination of large-scale deforestation and the continuous growing of barley on irrigated land turned huge areas into desert and scrubland, as well as exhausting the land and creating such high levels of salt that the land could no longer grow crops.

In Greece, deforestation and unsustainable farming resulted in denuded soils and scrubland on the hillsides. This story continued in the Roman period; the forests of Italy were all but wiped out by the increasing demand for fuel to heat public baths and to smelt metals. Home farming became more and more unproductive, owing to siltation, soil exhaustion and salination, as well as decreased rainfall caused by the deforestation. To overcome these problems, Roman leaders captured surrounding countries to supply the fuel, food and minerals to sustain the ever-voracious Roman Empire. In captured lands, such as North Africa, the land was farmed to exhaustion to feed Rome. Much of northern Africa, which had been fertile, ended up as just an extension of the Sahara Desert through the continuous growing of wheat and barley to fill Roman granaries.[15]

The Medieval European Vision

In the thirteenth century Thomas Aquinas combined Aristotle's comprehensive system of Nature with Christian theology and ethics. This was the conceptual framework that remained unquestioned throughout the medieval period. The Aristotelian vision of the world continued to dominate Christian thinking in Europe for fifteen centuries after Christ.

This vision was of a magical and organic view of Nature, in which the universe was knitted together by a web of correspondences that linked human nature, and the fate of individual men and women, to the wider cosmos. Man and Nature (or the microcosm and the macrocosm, as they were often called) were intimately bound.[16] The Florentine scholar Giovanni Pico della Mirandola (1463–1494) wrote: "God the craftsman blended our souls in the same mixing bowl with the celestial souls and of the same elements."

There were, however, some major shifts of thinking. The harmonious universe with its built-in order, as perceived by Aristotle, was replaced by a universe where the order came from God the creator, which could be disturbed by immorality or ignorance of God's plan. Through Grace we could understand God's plan and recreate that order. The goal of medieval science was to try to understand Nature—through a mixture of both reason and faith, with the purpose of understanding the meaning and significance of God's work. At this stage, there was no interest in the prediction and the control of Nature, which would have been seen as presumptuous to say the least.[17]

St Augustine

St Augustine (354–430) also played a part in developing ideas that still underlie much of the way we perceive and relate to Nature today. He was vital in formulating European ideas about both our 'nature' and, by implication, Nature itself. A repugnance of the sexual act (carnal lust) permeated Augustine's works. He created the concept of 'original sin': that Adam was created perfect by God and then was seduced by Eve—a woman, of course, who had been corrupted by the Devil—so that we are all of necessity born sinful because we were created both by the sinful carnal acts of Adam and Eve, from whom we are all descended, and by the carnal lust of our parents. The implication is therefore that our 'nature' is tainted, corrupted, imperfect and defiled. As a result, what was required was the aid of grace in order to cleanse our natural sin. It was at this point in history that Nature was beginning to be seen as both corrupted and female.

Pelagius, on the other hand, believed that sinning was not inherent in Man, on the grounds that Nature is inclusive and therefore also includes the possibility of avoiding sin. Unfortunately Augustine's ideas took hold but, as many have commented since, to question and to find fault in certain aspects of God's creation is to question not only His judgment, but also Nature, and this is the same as doubting our own nature.

The Transition

The Renaissance

The Renaissance was the transitional period between the Middle Ages and the Age of Enlightenment, roughly between the middle of the fifteenth century and the Reformation and Counter-reformation. It was a deeply religious age, but there was a new emphasis on worldly accomplishments and human abilities. The Greek philosopher Protagoras's aphorism that "Man is the measure of all things" became its motif.[18] One of the most important shifts in perception between the medieval perception of Man and his place in creation, and the Renaissance view, was the new glorification of human abilities, placing humanity at the centre of philosophical debate with its ability to understand Nature through its unique power of reason. In the subsequent Age of Enlightenment and the parallel Scientific Revolution, the medieval view of an organic, living and spiritual universe was beginning to be replaced by the idea of the world and Nature as 'machine'. Kenan Malik, in *Man, Beast and Zombie*, expressed it thus:

> "There were also the first stirrings of a new view of nature as an autonomous entity that proceeded according to its own laws without any external interference, a view that eventually gave rise to the scientific revolution. There was a growing awareness of natural order and a determination to see how far natural principles of causation—as opposed to divine intervention—could go in providing a satisfactory explanation of the world."[19]

Francis Bacon

Two figures stand out in the transitional period between medieval and modern thought in England, or more specifically between the Renaissance and the Age of Enlightenment: Francis Bacon (1561–1626) and Thomas Hobbes (1588–1679).

Bacon is recognized as the 'father' of the 'new philosophy', or modern science as we now call it. His attitude to Nature is revealing, in that it set the tone for the dawn of the modern age. He no longer saw Nature as 'mother', as self-regenerative, self-organizing and sustaining, but as 'passive-female' to be conquered by—of course—Man. The relationship was no longer one of reverence, but of domination. As Vandana Shiva has commented:

> "The removal of animistic, organic assumptions about the cosmos, constituted the death of nature—the most far-reaching effects of the scientific revolution. Because nature was now viewed as a system of dead, inert particles moved by external (rather than inherent) forces, the mechanical framework itself could legitimize the manipulation of nature."[20]

Bacon criticized Aristotelian philosophy for having "left Nature herself untouched and inviolate" and having sought just to "catch and grasp" her, rather than "seize and detain her".[21] As Vandana Shiva has said:

> "The scientific revolution was based on the destruction of concepts of a self-regenerative, self-organizing nature, which sustained all life. For Francis Bacon . . . nature was no longer a mother, but rather a female, to be conquered by an aggressive masculine mind." [22]

And if this critique seems a little harsh, let Bacon speak for himself:

> "Nature, in his view, had to be 'hounded in her wanderings', 'bound into service', and made a 'slave'. She was to be 'put in constraint', and the aim of the scientist was to 'torture nature's secrets from her'." [23]

And in his *Novum Organum*, Bacon wrote: "I am come in very truth leading to you Nature with all her children to bind her to your service and make her your slave."

Thomas Hobbes

Hobbes was one of the first philosophers to perceive Nature as wild and antipathetic to human civilization. He was convinced that without the structure of law and good government, we would all revert back to an unruly, wild state, permanently at war with each other. He perceived Nature as wild, brutal and untamed. There was human civilization on the one hand, and Nature and barbarism on the other; good 'civil' behaviour, under the rule of law as opposed to Nature, which was the product of random events, albeit based on natural laws. He believed, therefore, that we should not romanticize the condition of human beings in a state of nature, because we have thankfully superseded it.[24]

Jean Jacques Rousseau

These new rationalist ideas increasingly saw the workings of Nature and the universe as mechanistic and without meaning, but at every stage these ideas were being challenged. The Romantics who came after Hobbes rejected the idea of Nature as wild, untamed and uncivilized. This view was taken by John Milton (1608–1674) in *Paradise Lost*, in which he described the natural state enjoyed by Adam and Eve living in God's perfect garden. These ideas were taken to their ultimate conclusion by Jean Jacques Rousseau (1712–78). He believed in the idea of the 'noble savage', and rejected St Augustine's idea of original sin, taking the view that human beings are naturally good. He looked to inequalities in society for the causes of bad behaviour, believing that it was the structure of society and the hierarchical, oppressive system of government that was truly 'brutal'. He imagined the human race in a state of 'nature' before property and inequality existed. This was a re-creation of the myths of the Garden of Eden and the Golden Age. He imagined the noble savage as a vegetarian, happily going around eating berries and living the natural life, never subject to depression.

On an individual level, the equivalent state of the noble savage is childhood. So those two ideas led to a rejection of society and government and all that goes with it, and a return to the natural world on the one hand and to childhood on the other. These were the twin pillars of the Romantic movement. In the aesthetic hierarchy, Nature took on even more importance than art.[25] From my perspective, these ideas are about as unrealistic and ridiculous as the idea of Nature as wild, brutal and barbaric and in need of taming.

The Age of Enlightenment

In the eighteenth century there was a unique set of historical and social circumstances in Europe. Europe was in the process of change, a process that not only led to the worst aspects of the French revolution but at the same time to the creation of a dynamic atmosphere, in which new and creative ideas flourished in abundance. Everything was up for discussion. Every accepted idea was questioned. Nothing was sacrosanct. Reason, uncompromising logic and cutting analysis were the tools used in uncovering the 'truth'. Everywhere, emphasis was laid on the need for tangible evidence and clarity of thought. There was at the same time a movement away from anything that smelled of superstition, and metaphysics.

From this period on there was a growing sense of self-assurance, a belief that humans had the capacity to be truly the masters of our own destiny, without divine intervention, not only to control Nature by understanding how it functions, but to improve on it. There was no God-given order to the universe; it was we who had to create it.[26]

These ideas led many scientific minds to reject God altogether, but others, like Sir Isaac Newton (1642–1727), remained deeply religious, although they saw Nature as autonomous. Newton saw his task as revealing the workings of God's great work, Nature. He felt that if he could understand the workings of Nature and its natural laws, he would understand God's mind; indeed, as a mathematician himself, he was convinced that God himself must be the 'Great Mathematician'.[27] Newton, along with René

Descartes, was the father of modern science. It was he who discovered the immutable laws of Nature, describing them mathematically. He showed that mathematics is the basis of all the sciences. Without it there can be no real 'proofs' in science. Western philosophy needs reason and intellect, but science needs mathematics to 'prove' natural laws.

The Scientific Revolution

In understanding how attitudes to Nature changed and developed through the seventeenth century, the Age of Enlightenment, the Victorian period and on to the present day, we have to understand something of the methodology and paradigms of science. The father of scientific methodology was René Descartes (1596–1650). His scientific methodology was described in his *Discourse on the Method of Properly Guiding the Reason in the Search for Truth in the Sciences*. His goal was to enable people to become "masters and possessors of nature".

Here then was a fundamental change in the way the 'new scientists' saw nature. In the medieval world, the Church understood humankind as living in harmony with Nature, and the natural world as a gift from God to be revered and protected. The new philosophy, on the other hand, perceived Nature as having been created for our benefit alone and therefore an object to be exploited.

Out with the Old, in with the New

One of Francis Bacon's defining principles was that it was not enough to merely augment ideas that had gone before. "The entire work of the understanding must be begun afresh, and the mind itself be, from the start, not left to take its own course, but be guided step by step." This 'out with the old, in with the new' philosophy is very much part of the ethos of the modernist scientific approach, which has had very large repercussions for agriculture around the world. Instead of adapting and incorporating traditional forms of farming with the new methods, there has been, in many instances, a complete rejection of traditional practices and their wholesale replacement with modern methods, often with disastrous results.

The legacy of the 'Green Revolution' in the Punjab and Haryana regions of India, which began in the late 1960s and early 1970s, is a case in point. Traditional crops and methods were replaced with farm machinery, pesticides and fertilizers, as well as modern high-yielding varieties. To begin with, yields increased dramatically, but now the land is increasingly unable to support the burden of intensive agriculture. Crop yields and water resources are declining alarmingly, and some areas are almost barren. Many farmers are heavily in debt from their investments in new equipment and reliance on chemicals. Rural unemployment is increasing and there are ominous signs of a deteriorating farm economy.[28]

In the seventeenth and eighteenth centuries this new scientific approach replaced medieval methods (which were seen as muddled, unsystematic and superstitious) with a truly scientific methodology, based on a set of simple principles which could be followed by all, and verified and replicated by all. The purpose of these methods was to lead to a position where it was no longer possible to doubt or deny a truth, because through a systematic practice of 'methodical doubt' all uncorroborated evidence was dismissed, to be left with only that which was substantial and had survived all tests.[29] Jules Pretty, in his brilliantly researched book, *Regenerating Agriculture*, summarizes this approach as follows:

> "Since the early seventeenth century, scientific investigation has come to be dominated by the Cartesian paradigm, commonly called positivism or rationalism. This posits that there exists an objective external reality driven by immutable laws. Science seeks to discover the true nature of this reality, the ultimate aim being to discover, predict and control natural phenomena. Investigators proceed in the belief that they are detached from the world. The process of reductionism involves breaking down components of a complex world into discrete parts, analysing them and then making predictions about the world based on interpretations of these parts. Knowledge about the world is then summarized in the form of universal, or time-free and context-free generalizations or laws. The consequence is that investigation with a high degree of control over the system being studied has become equated with good science. And such science is equated with 'true' knowledge."[30]

However much one may applaud the scientific approach, it is not immune from criticism. Once a set of truths has generally been arrived at — often over-simplified, as in the plant nutrition debate—they begin to take on a momentum of their own. Over time a complex set of ideas develops, which become increasingly accepted orthodoxy. Research which supports the orthodoxy is more likely to be undertaken, and the results adopted. Conversely ideas which do not fit into the paradigm are less likely to be supported or accepted.³¹ This is precisely the story of conventional farming theory and practice, as it has developed over the last two centuries.

Justus von Liebig

In the atmosphere of the Age of Enlightenment, the Scientific Revolution was inevitable. Humphrey Davy (1778–1829) undertook the chemical analysis of both manures and plants, which gave a clue to the existence of elements common to both, but it was Justus von Liebig (1803–73) and other chemists who calculated the amounts of different chemicals in different crops.

Liebig was a typical product of his age. He burned plants and then analysed the remaining ash, concluding that the minerals that were left were those that plants needed for good growth. The three main chemicals he identified were nitrogen, phosphorus and potash, known in the trade as NPK. The theory up until then had been that plants fed on humus. To test this, Liebig left some humus in water for some time. He then filtered and dried it. He found no minerals left and concluded that humus was useless as a plant food, which of course it is on its own. What he did not know was that the soil was alive with millions of micro-organisms that were able to convert nutrients from this 'store' of humus into available plant food. Nonetheless, his discovery of the three basic essential water-soluble plant foods was of great importance, as was with his discovery that if any one of the main plant nutrients is deficient, then the growth of the plant is restricted, however much of the other nutrients is present, the so called 'Law of the Minimum'. Another of his vital discoveries was that it is the carbon dioxide in the air and not the carbon in humus that is the source of carbon in plants.³²

Later in life, Liebig was very critical of the developments that took place as a result of his work, the oversimplification of his findings, and the way crop nutrition had been abstracted from its ecological and biological context. The history of science is full of lesser intellects oversimplifying the original ideas of its great pioneers. This limited understanding of soil ecology and plant nutrition still plagues agricultural science and practice to this day. Starting from this oversimplified vision of plant nutrition divorced from its ecological and biological context, a huge industry was built.

Paternalism versus Mother Nature

There is another factor that has to be taken into account in this story, which we have already touched on: that of the increasingly paternalistic attitude, with 'Man' as creator and Nature as something that needed to be 'improved'. This new, mechanical view of the universe perceived the soil as inert, no longer as *terra mater* ('mother earth'). Mother Nature was replaced with a new paternalistic paradigm. As Vandana Shiva has said, "The construct of the inert earth was given a new and sinister significance, as development denied the earth's productive capacity and created systems of agriculture that could not regenerate or sustain themselves." [33]

In other words, Nature was no longer trusted as capable of organizing and regenerating itself. In many ways, this lack of trust in Nature stemmed from the experiences of history—diseases and plagues that people felt powerless to deal with. However, in a lot of historical cases, the causes were induced by people themselves, albeit through ignorance and exploitation. For example, the Irish potato famine of 1846–7 was brought about by rural poverty, bad landlords and overcrowding, resulting in the over-dependence on one crop.

The idea that the soil/plant complex contained all the biological mechanisms to grow and sustain plants and animals gave way to the idea that people had to add things to it to help plants to grow. Vandana Shiva has shown how the paradigm behind the 'Green Revolution' substituted the regenerative nutrient cycle inherent in traditional sustainable agriculture "with linear flows of purchased inputs of chemical fertilizers from factories and marketed outputs of agricultural commodities." [34]

The sad thing is that this became a self-fulfilling prophecy. Using chemical fertilizers while neglecting humus creation caused the soil life to die out, so in order to grow crops at all, chemical fertilizers became a necessity.

When trying to understand existing scientific orthodoxy, one has to realize that 'investigators proceed in the belief that they are detached from the world'. However objective the investigative method is, the scientific facts arrived at have to be interpreted and understood in context. This inevitably involves the subjective interpretation of the scientist, which is coloured by the belief systems and paradigms to which he or she has been subjected. As Kenan Malik has commented, in the seventeenth century, science and philosophy were originally one body of knowledge called Natural Philosophy, but by the eighteenth century a divergence had developed: "The separation of science and philosophy meant that scientists . . . could remain blind to the philosophical assumptions that animated their work, and at the same time pass off philosophical speculation as fact." [35]

The Birth of Fertilizers

The idea that Nature was not self-sustaining started to be reflected at the practical level. In the mid-nineteenth century, the largest and most 'progressive' farms were already starting to use imported dried Peruvian guano and oilcake (the by-product of vegetable oil production) as a nitrogen fertilizer, and powdered bones (calcium phosphate) or Lawes's superphosphate, which was made by treating powdered bones or mineral phosphates with sulphuric acid, although for most farmers farmyard manure, clover breaks and rotation remained the main forms of fertilization. But a trend had been started towards 'improving' Nature through a simplified and systematic scientific system of inorganic husbandry.

In its most reductive form, this means single crops being grown year after year and fed solely with chemical fertilizers, with pests and diseases controlled exclusively with pesticides. Indeed, monocropping cannot be sustained without the use of pesticides, just as intensive animal rearing cannot function without the daily use of antibiotics. In this system where cereal crops are grown year after year, the soil becomes a largely lifeless medium to prop up the plants and act as a not very good sponge to retain water.

New Discoveries in Soil Science

Interestingly, despite this growing trend, new discoveries in soil science began to transform the way biologists, at least, looked at plant nutrition: they saw a much more fascinating and complicated story. Towards the end of the nineteenth century, the importance of the nitrogen-fixing bacteria in the root nodules of leguminous crops in the rotation was at last understood. The value of legumes had been known for hundreds of years in Europe and for even longer in countries like China, but the scientific understanding of the mechanism involved led to even more important discoveries about soil ecology and plant nutrition.

The concept that soils were just an amalgam of chemical compounds became inadequate as an explanation. This was the beginning of the recognition that the soil was alive with bacteria and other micro-organisms which had an influence both on the structure of the soil and on plant nutrition. Not only was it the beginning of a fuller understanding of soil ecology and plant nutrition, but it became one of the main inspirations for the developing ideas of the early pioneers of the organic movement, such as Sir Albert Howard. However, the further developments of these ideas were not generally taken up or explored, for the most part, by the agricultural mainstream.

The Industrial and Agricultural Revolutions

The alienation from Nature was further exacerbated by the industrial and agricultural revolutions of the eighteenth and nineteenth centuries, when the majority of the population moved from the countryside into the new and expanding industrial towns and cities. The idea of Nature as 'out there' took on a new and poignant reality. At the same time, the final stages of the enclosure of land and the expropriation of common land was taking place. There was a feeling of connectedness with common land by local people, and a personal connection with Nature. Although there remains 500,000 hectares (1.25 million acres) of common land in England and Wales, this is minuscule when compared with the area before the beginning of the eighteenth century, where every village had its local common land, both for grazing and for recreation.[36]

At the same time, industrial developments were inexorably changing agricultural practice, with the increasing use of machines such as the threshing machine, run by the new steam traction engines, and later steam-driven ploughs and horse-drawn cutters and binders. With the repeal of the corn laws in 1846, which had protected British farmers against cheaper imported wheat, and the population explosion that was occurring, the drive was on to make British agriculture more 'economic' by reducing the number of farm labourers, and replacing them with machines, and by growing more and more specialist cash crops.

The Romantic Reaction

This increasing alienation from Nature produced an inevitable reaction against the perceived mechanical and unspiritual vision of science and technology, which the Enlightenment and the harsher aspects of the Industrial Revolution brought about. This reaction manifested itself in the form of the Romantic Movement.

Immanuel Kant (1724–1804)

The German philosopher Immanuel Kant was very much influenced by Rousseau in his aesthetic theory. Kant's view was that the experience of beauty has as its primary object the natural world, and that Nature is higher in aesthetic values than art because it is something towards which we can have a completely disinterested attitude. We stand back and contemplate it, and in contemplating it we have a kind of religious epiphany in which we see revealed not just the meaning of the world, but our own harmony with it. These ideas, along with Rousseau's view of Nature as the fount of innocence, had a profound influence at the time. Kant's aesthetic vision also had a profound influence on Schiller and other German Romantic poets and the German Romantic painters.[37]

The English Romantic Poets

The reaction against the Age of Enlightenment and the Industrial Revolution was epitomized by William Blake (1757–1827), and later the Romantic poets Coleridge (1772–1834) and Wordsworth (1770–1850). As William Blake saw it, innocence was something that had to be cultivated. He, above all others, was the most sceptical about the logical and scientific approach to understanding, as witness his famous painting of Sir Isaac Newton turning his back on the very Nature he was trying to understand through mathematics and logic. Blake loved Nature and intensely disliked the dehumanizing effects of the Industrial Revolution and its 'dark Satanic mills'.

Coleridge and Wordsworth, in their turn, were greatly influenced by the ideas of Rousseau and Kant, and indeed Coleridge translated Kant and introduced his ideas to Britain. At the beginning of the eighteenth century, for instance, mountains were seen as an inconvenience for the traveller, or somewhere only frequented by shepherds. But as urbanization and industrialization took off in the latter half of the eighteenth century, people felt an increasing sense of alienation of the human spirit. In reaction they tried to get out into the countryside when they could, and began to see mountains in new ways. Writers like Wordsworth and Coleridge now had a spiritual language for talking about Nature. It was as if Nature had been disenchanted, and its magic had been removed by the scientific revolution, and they attempted to overturn this process through their poetry.[38]

Many in the Romantic Movement, however, still saw Nature as 'other'. Although they admired, took inspiration from and in some cases almost worshipped it, it was still perceived as separate—something that exemplified the naturalness and innocence that we have lost. Wordsworth was somewhat exceptional in perceiving contact with Nature as a way of making contact with one's true Self, and recognizing both as the same, as in his lines composed at Tintern Abbey:

> Once again
> Do I behold these steep and lofty cliffs,
> That on a wild secluded scene impress
> Thoughts of more deep seclusion, and connect.

And later:

> And I have felt
> A presence that disturbs me with the joy
> Of elevated thoughts; a sense sublime
> Of something far more deeply interfused,
> Whose dwelling is the light of setting suns,
> And the round ocean and the living air,
> And the blue sky, and the mind of man:
> A motion and a spirit, that impels
> All thinking things, all objects of all thought,
> And rolls through all things. Therefore
> I am still.

It is clear from lines like these that Wordsworth knew from his own experience that in order to truly commune with Nature you have first to be in touch with your own 'nature', which is in essence the same. Nature can be the trigger, but without that developed inward sensitivity, the connection cannot be made and there will be no recognition.

Charles Darwin (1809–1882)

Charles Darwin rewrote our place in Nature in a radical way, a way that people today are still trying to come to terms with, in his assertion that all life on earth has one common ancestor, and that humans are only one of the most recent in a long line stretching back to a single-celled creature at the beginning of life. Nature could no longer be seen as something static; it was constantly evolving and changing in response to its own inner movement. For many, Darwin's theory of evolution is the final proof that human beings are an integral part of the natural world. However, his theories have been interpreted in many ways, according to the beliefs and paradigms of the time in which they were viewed.

Age of Enlightenment ideas about the uniqueness of the human intellect, religious beliefs about humanity's special place, and the Victorian conviction of superiority of European culture over others, inevitably affected interpretations of Darwin's ideas. He himself spoke of 'lower' and 'higher' life forms, but he nevertheless believed that the differences exhibited by humans were not unique to us, just more developed than is seen in animals. He wrote:

"The difference in mind between man and the higher animals, great as it is, is certainly one of degree and not of kind. . . . The senses and the intuitions, the various emotion and faculties, such as love, memory, attention, curiosity, imitation, reason, etc., of which man boasts, may be found in an incipient, or even sometimes in a well-developed condition, in the lower animals.[39]

Darwin's theory of the survival of the fittest was very popular in the late nineteenth and early twentieth century climate of unbridled capitalism and imperialism, and was used to justify both. Here, Nature was seen as 'red in tooth and claw', with all the species fighting and competing with each other for survival. However, there are many examples, as in the life and ecology of the soil, where different species are seen to have evolved together, mutually benefiting each other—as when a plant's root excretions feed bacteria around the root hairs, stimulating them to make nutrients available for the plant. This recognition of species 'co-operation' was described by such writers as Kropotkin, a late nineteenth-century anarchist writer. In his book called *Mutual Aid*, he used examples in Darwin's writings of species actually relying on each other and often helping each other.

Darwin's ideas have also been used to bolster the Renaissance and Age of Enlightenment beliefs in the continuous progress of evolution which still colours much of our beliefs today. Indeed Darwin himself accepted the idea of progress in evolution.[40] It could be argued that the idea of continuous progress is directly due to perceiving Nature as imperfect at every stage of its development, which then led to the idea that it is our duty to help perfect that which is flawed. Most interestingly, Darwinists today have discarded the progressive view of evolution, believing that it does not work towards any goal. However, the deeply held conviction that continuous progress is both inevitable and healthy lies beneath much of Western thinking, including the conventional approach to farming. With modern farming, as with many other fields of endeavour in the modern world, a specific line of development, once started, is mistakenly seen as the only route to progress.

Linear and Cyclical Perceptions of Reality

The idea of continuous progress, from our primitive roots to increasing states of civilization, is a peculiarly linear, Western view. Although not all Eastern traditional beliefs are the same, there is nonetheless a theme that runs through most of them: that of events being cyclical. This has had as much of an effect on Eastern views of Nature as the linear model has had on the Western view.

In Eastern cultures, this cyclical view of Nature created a climate where organic farming practices were natural and obvious. The seasons and the phases of the moon were cyclical. Civilizations were seen to grow and decline, societies to pass from periods of ignorance to enlightenment and back again. Even the universe itself was seen as cyclical, from creation to dissolution to creation.[41] The belief in the cycle of birth, death and rebirth was reflected directly in the farmer's recognition of the death and breakdown of plants and animal products into compost, giving life to the next generation of plants and animals. With farmers immersed in a culture where cyclical ideas were part of everyday understanding, it was inevitable that sustainable agriculture, with its serious recycling of organic waste, was practised over thousands of years. It was not until Western science and technology were adopted, and in China, European Marxism and the wholesale rejection of the old beliefs, that ideas of continuous progress and modern conventional farming took hold.

Darwin has without doubt left a huge legacy, and despite differences of interpretation, he changed our ideas about our relation to Nature for ever. In many ways he has rekindled more ancient beliefs, with their emphasis on humans as an integral part of Nature, albeit from a different perspective.

The Twentieth and Twenty-First Centuries

The destruction of the land by unsustainable forms of farming continued into the early twentieth century in the southern and south-western states of America. Farmers were being encouraged to expand westward by the Government, and the lure of the high price of wheat at the time. In the first decade, 30,000 farmers registered holdings every year. In 1919 alone, some 4.5 million hectares (over 11 million acres) of grassland were ploughed up for the first time to grow wheat. In the 1930s the fine sandy topsoil, denuded of its humus, blew away, creating a huge dustbowl that destroyed not only the land but the farmers, their families and whole communities who were trying to eke out a living.

The First World War brought about the large-scale production of, among other things, nitrogen fertilizers. Large industrial chemical companies in countries like Britain, Germany and the USA invested huge amounts of money in the production of explosives to meet the new demand. After the war, to maintain their hugely increased production capacity, they promoted ammonium sulphate along with phosphate and potassium fertilizers on a scale hitherto unseen. The Second World War cemented this trend, with a huge push in the UK to be as self-sufficient in food production as possible. With a real threat of starvation, all the stops were pulled out. This led to the establishment of the modern fertilizer and pesticide industries and a period of unparalleled use of chemicals that continues around the world to this day.

It was the arsenal of modern pesticides that finally completed the picture of modern chemical farming from 1945 onwards— and I use the word 'arsenal' deliberately, because the marketing and advertising strategies of companies that manufacture and sell pesticides often sound like those of arms dealers. Their products have names such as Assassin, Avenge, Commando, Crusader, Missile, etc. It is all about making war on the enemy: Nature. These same companies often have a dark history, because they have also developed chemicals for use in real wars. Dow Chemicals invented napalm, a particularly pernicious substance which led to many crops—and civilians—being burned in the Vietnam War. Monsanto invented Agent Orange, a defoliant which was also used in Vietnam to destroy crops and forests, and which has led to birth defects in children and other problems.

This is the vision of total warfare, whether on Nature or on human beings: chemicals to feed crops, chemicals to control pests and chemicals to administer to animals. As a result we have become insensitive to the damage we are doing to our own environment, the fertility of the soil and indeed to our own health, accepting it as 'the price of progress'.

Chapter 8

The Challenges Ahead

"An important step is the radical reform of national agricultural policies. Without such change, advances seen to date will stay small-scale and parochial."—Jules Pretty[1]

Global Warming and the Energy Crisis

When discussing organic and sustainable agriculture, one cannot avoid the subject of the limited life of fossil fuels and global warming. To quote the Soil Association's response to DEFRA's Policy Commission on the Future of Farming and Food:

> "No current agricultural systems are sustainable. The use of fossil fuels is not sustainable. Farming where fertility and crop protection are based on any input of artificial fertilizers and pesticides is not sustainable. Organic farming is not sustainable either. What organic farming can reasonably claim is that the system is based on achieving and maintaining both soil fertility, and healthy crops and farm animals, through mechanisms which are currently more sustainable than any other system, and which have the potential to be truly sustainable in future. Organic farming is based on values which embrace the concept of sustainability, however imperfect current practice is.[2]

When discussing fossil fuels and global warming, there is often a general lack of awareness about how immediate the problems are—how the serious changes to the climate caused by global warming and the scarce resources of oil and natural gas will result in a dramatic change in the direction society will have to take to cope with these new realities. As Mayer Hillman has said in his book *How Can We Save the Planet*:

"The apparent contradiction between belief and action may be because there is a feeling that climate change will turn out to be an ephemeral problem that will magically disappear or that human ingenuity will lead to the development of some miraculous technology so that our material standards of living can continue for ever. We challenge both these convenient myths."[3]

If we used up all our fossil fuel supplies, it would have the worst possible effects on global warming, so we need to reduce our dependence on fossil fuels dramatically as soon as possible.[4] There have been many informed calculations about the amount of fuel we have left, but according to the most optimistic estimates of the US Department of Energy, it will only last until 2037 at current rates of consumption![5] In making this calculation, the oil companies have factored in data about sources that have not yet been tapped, including digital X-rays from satellites, seismic data and 641 exploratory wells.[6]

However, there is one telling part of the projection that is often overlooked: the phrase 'at current rates of consumption'. But the evidence shows that the consumption of oil is rising at around 2.8 per cent per year. This then reduces the time left even further, to around 2030. Then we have to take into the account the growth of the world's population which is growing at a rate of a billion every twelve years, whilst at the same time China and other countries like India and Mexico are busily industrializing and will require even more of the 'cake'. It would also be naive in the extreme to think that oil prices and availability will remain the same. At the time of printing this book, prices have already begun to rise steeply. Some estimates say oil production peaked around 2000, others that it peaked in 2004, and yet others that it will do so in 2010.[7]

However, with consumption increasing faster each year, the peak in supplies has probably already arrived, with all its attendant consequences. What is required is an urgent switch to alternatives to oil, particularly fuel-cell technology, combined with the development of renewable energy sources as quickly as possible. Moreover, these new technologies must be combined with both savings in energy and dramatic increases in its efficient use.

Probably the most powerful argument for both the necessity and practicality of transferring to renewable energy sources, is presented by Hermann Scheer in his book *A Solar Manifesto*. He is one of the world's most knowledgeable experts on the subject, a member of the German parliament and President of EUROSOLAR, the European Organization for Renewable Energies.

So how does this fit in with our discussion about organic farming? Vast amounts of energy are used to manufacture fertilizers. It takes around five tonnes of oil, or its equivalent, to produce one tonne of chemical fertilizer. The raw materials have to be mined, using oil, then have to be converted into fertilizers, using more oil, then transported to the farms—using yet more oil. Intensive animal production systems require huge amounts of oil for the manufacture of buildings and their heating. The fact is that modern agriculture, on average, uses more energy in producing the food than is obtained from the crops themselves. Organic production, on the other hand, has been shown by several studies to require up to 60 per cent less fossil energy per unit of food produced, despite the extra energy required for weed control.[8] Moreover, with the ever-increasing use of no-tillage techniques, this figure should come down even more.

Measuring energy use is one of the best ways to measure the true productivity and efficiency of any system of agriculture, as well as how ecological its production methods really are. Nicolas Lampkin has compared the energy efficiency of conventional and organic systems of farming in his book *Organic Farming*:

> "Conventional systems are not sustainable given their physical, chemical and biological impacts on the soil, their excessive consumption of non-renewable resources and their far-reaching effects on the global ecosystem. The only farming systems developed today, which offer a significant degree of sustainability in respect of all these resources, all have a basis in the organic model of carefully designed crop rotations, maximal internal cycling of nutrients, and enhancement of, rather than substitution for, natural biological processes."[9]

Population

In considering the problem of feeding the world using sustainable or conventional systems of farming, one has to look at the challenge of the current population explosion. The discussion usually revolves around how to grow more and more food, but this is completely inadequate as an answer to the problem. To believe that the population is going to continue to grow exponentially *ad infinitum* is to misunderstand the laws of physics and biology. In the 1900s there were over 5.6 hectares (14 acres) of productive agricultural land per person on the planet; today there are only 1.5 hectares (3.7 acres), only 0.4 hectares (1 acre) of which is arable.

In the 1960s and 1970s there was an attempt to tackle the population explosion with programmes to provide easily available birth control, especially in the Third World. That appears to have fallen by the wayside. Yet any policy on feeding this world must include population control, otherwise starvation and increasing disease will control it for us. In fact it is already happening. At the beginning of the twenty-first century, 850 million people are hungry every day, and more than 15 million—most of them children—die of starvation each year. In 1960 there were 3 billion people on the planet. Fourteen years later, in 1974, there were 4 billion. Thirteen years later, in 1987, it was 5 billion, and in 1999, 6 billion![10] Admittedly, since then population growth has slowed, but even so it is estimated that it will reach 7 billion in 2013.

This cannot continue. Even without the fossil fuel crisis, this exponential growth rate would soon crash. With the impending fuel crisis, the crash will come even sooner. There are countries where the population is declining, but they are usually already some of the most heavily populated, such as Japan and some European countries. In the rest of the world, the story is very different. Over 100 countries are now net food importers. In the late 1990s, the Washington-based Population Institute published a report which documented eighty-two of these as unable even to buy enough food for their populations.[11] At the same time only a few dozen are exporters—largely Canada and the USA—and they export less than 230 million tonnes of grain per year. This will inevitably decrease, whilst demand will increase. This combination of fossil fuel decline and population explosion, should be enough to encourage administrations around the world to change direction now.

The Future for Western Agriculture

For those Western countries which have taken the modern chemical industrial route for agriculture, the change of direction required in the near future will involve a major upheaval. The complex belief systems and commercial and financial structures are like vast tankers: they will take a very long time to turn around. There is no doubt it is possible, indeed essential, but the main resistance to it comes not so much from the physical infrastructure and the denuded soils and landscapes, but from the many professional careers that have been built on the conventional paradigm. The change in direction is going to be hardest for the orthodox agricultural, political and scientific establishment.

One would imagine that educational establishments would be places buzzing with new ideas, encouraged by enthusiastic professors. Unfortunately, with the exception of some notable establishments, the reverse is usually the case. Whole academic careers are built out of accepted theories, and among the farming community itself there are now a couple of generations who have only ever farmed in the conventional way, believing they were at the cutting edge of new agriculture. Most of these farmers have never seen healthy, vibrant living soil with high humus, high worm contents and good tilths. They are completely reliant on chemical imports, fossil fuels, large equipment and huge debts—which in the UK for example, now stand at £10 billion and rising.[12] However, moves are being made to change direction and face up to the problems. These changes vary from country to country; some, like Switzerland, are taking major steps towards sustainable farming, while others are lagging behind.[13] Changes are beginning to happen, but as yet there is little sign of major attempts by administrations to make the profound changes that are required in both ideas and agricultural production if we are to maintain soil fertility and feed a future world.

Renewable Energy Production

As the Soil Association itself has said, even organic farming, especially in industrial countries, is not truly sustainable, but it does provide a good basis for sustainable systems in the future. One of the ways this can be achieved is by producing renewable forms of energy on the farm itself. Renewable energy production is ideal for both conventional and organic farms. It is the kind of diversification that is perfect for farmers, who could save, and even make, money from such projects, as well as helping to cope with the impending fossil fuel crisis. The types of energy production involved include bio-gas production, wind power, water power, bio-energy crops and photo-voltaic electrical production.

Bio-gas production is one form of energy production for farms that have animals. As we have already seen, bio-gas can be used directly for heating, refrigeration, running vehicles and cooking, or it can produce electricity by highly efficient direct fuel-cell technology. In the near future it could be used for the production of synthetic diesel. There is obviously a great potential for farms with cattle and/or pigs. A bio-gas plant could be even more productive if it included other farms' slurry and local household sewage, where economies of scale make it more economic. To use large amounts of valuable organic matter in producing energy in this way is not the best use of manure to build soil fertility directly. However, the resultant by-product does not lose any of its original nitrogen. If this high-nitrogen slurry was used as a compost activator to rot down high-cellulose straw and other waste into valuable compost, or to grow green manure crops or grain crops with valuable high cellulose straw, to produce even more compost, then the building of soil humus would not be severely compromised. Bio-gas technology is well established now, with engineering companies who specialize in installing units.

Wind turbines are becoming more and more efficient and cost-effective, which is why wind power is increasingly popular all around the world on low-productivity upland hill farms. On a smaller scale it can be used to generate electricity for the farm itself in combination with batteries, or it can be fed into the national grid to save on fuel bills. Large-scale wind farms on and off shore are of course also essential, but smaller-scale projects can make a significant contribution and at the same time supplement the farmer's income.

Pelton wheels are the most efficient way of producing electricity from water on a small scale for those farms that have access to a steady flow. The water is fed by gravity through a venturi (a pipe which becomes increasingly restricted towards one end). This forces the water out at great speed, hitting the cups of the pelton wheel. These cups are designed to transfer the maximum amount of energy to the wheel, which drives an alternator.

Photovoltaic (PV) panels have flat silicon crystals that produce an electric current when the sun shines on them—just like a larger and more efficient version of a light-driven pocket calculator. The capital costs of the technology are still high, but they are rapidly decreasing, and farms are ideal sites, as they usually have plenty of roof and other space on which to place the panels. PV panels are ideally suited to run in combination with windmills, because wind is usually available when the sun is not, and vice versa.

Hydrogen is the fuel of the near future. All the renewable forms of electrical production—wind, water and photovoltaic—can be used to split water to produce hydrogen which can run fuel-cell-driven tractors, other vehicles and equipment. Many companies are now working on the development of extraordinarily efficient and powerful electric motors that are powered by hydrogen fuel cells.[14] Car companies already have working demonstration vehicles. Honda started marketing fuel-cell vehicles in 2002, and Toyota's fuel cell SUV, the Fine-N, is the first such vehicle to have acquired certification under Japan's Road Vehicle Act. Leases for the vehicle began on 1st July 2005, with commercial production by 2015. The hydrogen is passed through a battery-like fuel cell where it combines with air to produce electricity. This processes is so efficient that 50–60 per cent of the energy contained in the hydrogen is converted into electricity.

The stored hydrogen also acts as a superior battery, takes up less space than conventional batteries and offers long-term or large-scale energy storage. Windpower and PV-generated power are intermittent, but with an integrated system including an electrolyser (to split water into hydrogen and oxygen), a pressurized gas store plus fuel cells and a specially designed computer programme, an even supply can be maintained.[15]

There are many farms, especially those that are isolated, which might use the energy produced solely on the farm, but it is best linked into the National Grid. Unfortunately in the UK, the price paid by the power companies to producers of electricity is far too low. In some European countries, notably Germany, private producers get far more. So there is a need for government legislation to set decent tariffs.[16] There is also resistance from the companies to co-operating with smaller producers. To overcome these difficulties there need to be central initiatives to encourage and facilitate smaller electrical producers who wish to feed into the grid. There are of course technical problems, but these can be overcome, given the will to do so. Already farmers are able to tap into the national grid, with the costs of their production deducted from the farm's consumption. This is done by having both an incoming and an outgoing meter. Even if the farm only produces enough for its own needs, the grid ensures a steadiness of supply.

Certain energy crops, such as sugar beet, contain high levels of sugar which can be turned into alcohol to drive vehicles. In Brazil and other countries, sugar cane is used in just this way, and vehicles are increasingly run on bioethanol. Another type of energy crop is biomass, such as coppice woodland, where the trees are cut every few years and the wood used in power plants. This sort of technology is becoming popular in countries like Sweden. The advantage of wood as a fuel is that it can be grown where other agricultural crops cannot, and therefore does not encroach on food production. All renewable energy production has to be seen in the context of a possible major catastrophe caused by global warming, in which all our attempts at growing food, whether by organic or conventional means, would be severely restricted. And while most supporters of sustainable energy would also want see solar electric, wind, water and hydrogen as the main sources of power production, they would recognize that energy crops have a part to play. The founder of the organic Elm Farm Research Centre, David Astor, believed that the economic and social implications of approaching the limits of finite resources (especially oil) without credible technical, economic and social alternatives in place would be far more horrendous than anything civilized society has been confronted with to date.

In the UK Government's Strategy for Sustainable Farming and Food there is a section on encouraging non-food crops, particularly energy crops as an alternative to fossil fuels: "The government is working with partners to establish a non-food crops centre to drive forward innovation." This is one of the more exciting moves by the UK government, and this initiative is now up and running. The government has recognized that we will need to provide at least some of our sustainable energy from agricultural and forestry production. It goes on: "Defra and DTI (Department of Trade and Industry) are also considering the potential role of demonstration projects to increase commercial uptake of renewable raw materials, working with the key players in industry, research and technology transfer", and "Defra's Energy Crops Scheme provides grants towards the costs of establishing short-rotation coppice and miscanthus, and for establishing producer groups set up to supply short-rotation coppice to power stations and other energy uses." The Government will provide at least £66 million for this project, and there will also be a 20p (45 per cent) per litre reduction in the duty on bioethanol (a vehicle petrol substitute produced from such crops).[17]

There is an important caveat to this idea, however. It is essential not to use conventional methods to produce energy crops, otherwise it becomes self-defeating, because of the amount of energy needed to grow crops conventionally. For many years peat has been mined in Sweden and Ireland to burn in power stations. However, this is a one-way street, as huge areas are being denuded of peat, not to mention the ecological damage. As an alternative, there were plans in Sweden to plant fast-growing willows on the peat, with every fifth tree being an alder and the ground sown with perennial lupins, both of which have bacterial nodules on their roots which supply the willows with nitrogen and allow the regular cutting of wood without a loss of nitrogen from the soil. The willow and alder are harvested every few years, chipped and fed into the boilers at the power stations.[18] This form of production is sustainable, but there are other forms of temporary willow plantation that use sewage sludge.

Renewable energy production has the capacity to make organic farming truly sustainable, even capable of producing a net excess of energy both from food and direct energy production. Much emphasis has been placed by governments on large-scale projects, but the aim to supply a large percentage of the country's power from renewable sources will only be achieved if these larger projects are complemented by as many small on-site projects as possible. Substantial government grants are now available in the UK and other European countries to help offset capital costs when installing photovoltaics etc.[19]

Food Miles

The other huge use of energy used in food production and distribution (the food chain), is in the transport of food over sometimes vast distances (called food miles). The use of energy used in the handling and transport of food is estimated to account for 16–21 per cent of the UK's total energy bill. To put it another way, it has been estimated that the contents of an average supermarket trolley have travelled 3,000 miles to get to the checkout![20] To carry food unnecessarily around the country rather than selling local produce locally is unsustainable. One of the problems is the way food distribution networks are arranged. Supermarkets have too few large regional depots, to which the farmers have to transport their produce. It is all too common for food produced in a local area to travel half way across the country to a depot, only to be transported back and sold a few miles from where it was grown. It is part of the UK government's strategy to encourage the leading supermarkets to buy more local food, but the distribution structure needs overhauling. According to research by the Department for Environment, Food and Rural Affairs published in July 2005, the transport of food across the country and around the world is costing the UK £9 billion a year, and the average Briton travels 898 miles a year to shop for food. DEFRA has vowed to work with the food industry, aiming to reduce food transport by 20 per cent by 2012.

Another aspect of food miles is unnecessary trade. For instance in 2000, the UK imported 255,000 tonnes of pork and 94,000 tonnes of lamb, while exporting 195,000 tonnes of pork and 98,000 tonnes of lamb. This kind of idiocy and waste of energy has to be discouraged.[21]

One increasingly popular way to cut down on food miles is the sale of local food, whether through farmers' markets, box schemes or farm shops. In 2003 the sales of organic food through these outlets, direct from producers, was well over £90 million. As part of this trend, regional food is enjoying a revival in the UK. Purchasing more local food not only reduces fuel consumption but helps farmers, who are able to add value to their products through processing them, selling direct to the customer, or just obtaining a premium for local quality and individuality. And customers have the chance to obtain fresh local produce from suppliers and producers they know—what has been dubbed 'traceability'.

Of course much of this is just what we did in the past, albeit in new forms, but there is nothing wrong in admitting mistakes and rethinking our strategies. An imaginative modern approach to encouraging the sale of local produce is through projects like the website BigBarn.co.uk. By feeding in one's postcode one can get a list of local suppliers, farm shops, markets etc., and a map of how to get there.

Importing organic food that we could easily grow in the UK from other European countries, or even from halfway across the world, as we do, goes against the whole ethos of organic and sustainable farming. To cut down imports and food miles, we should aim to grow as much of our food as possible ourselves, and to import those items that can only be grown in other countries. What all countries should aim for is to be as self-sufficient as possible, while producing an excess of certain foods which are particularly suitable for that country's local climate, such as quality regional foods. These can then be traded and offset against those that have to be imported. The UK only produces 63 per cent of the food it consumes, but 57 per cent of our imports consist of foods that could be grown here, leaving only 43 per cent that can only be grown in other countries. This is at a time when our food imports stand at £18.2 billion per year and exports at £8.5 billion—a deficit of £9.7 billion.[22]

Many economists see no problem in relying so much on imports; some even believe that the UK could survive quite happily if it imported all its food, because food constitutes only about 8 per cent of our imports. Luckily this short-sighted and dangerous urban ignorance is not current government thinking; knowledgeable citizens throughout the ages have always understood that agriculture is the foundation of all societies. History is littered with the bones of civilizations that have fallen in the past because they ignored this most fundamental of facts.

Agricultural Efficiency and Productivity

Indeed, one of the things that needs to be challenged is the way Western economists judge the efficiency and productivity of industry and business in general, and agriculture in particular. It became obvious to many people more than twenty years ago that the orthodox ways of measuring these factors would have to be changed to accommodate a much broader ecological and social perspective. As Fritjof Capra made clear in his book *The Turning Point*:

> "Since the conceptual framework of economics is ill suited to account for the social and environmental costs generated by all economic activity, economists have tended to ignore these costs, labelling them 'external' variables that do not fit into their theoretical models. . . . To deal with economic phenomena from an ecological perspective, economists will have to revise their basic concepts in a drastic way. Since most of these concepts were narrowly defined and have been used without their social and ecological context, they are no longer appropriate to map economic activities in our fundamentally interdependent world." [23]

A prime example of this is conventional farming in America, which is highly mechanized and petroleum-subsidized. Although it is regarded by some as highly efficient, it is in fact the most inefficient in the world when measured in terms of the amount of energy used for any given output of calories, not to mention the expensive pollution of the environment.[24]

One common way of measuring agricultural efficiency is to see it purely in terms of the production of a single crop per hectare, regardless of the input of energy required for every extra tonne produced, and the incremental damage to the environment. It is obvious that if one only takes a limited number of parameters into account when calculating efficiency, the results become meaningless. When a farm is run with a high level of biodiversity, both in the variety of crops and animals, and in the diversity of habitats, a different model of efficiency and economics prevails, in which each part supports all the others, both ecologically and financially. For example, the native farmers in Chiapas, Mexico are seen as backward because they produce only two tonnes of maize per hectare, instead of the six tonnes grown on modern Mexican farms. However, these native farmers also grow beans up the maize stalks, fruit, pumpkins, sweet potatoes, tomatoes, a variety of vegetables and medicinal herbs—all on the same plot of land. In all, 15 tonnes of food are produced per hectare (6 tonnes per acre) all organically, without financial aids, pesticides and fertilizers.[25]

Vandana Shiva is one of the more recent proponents of a complete reappraisal of our economic parameters. She commented in her book *Biopiracy*:

> "We need a transition to an alternative economic paradigm that does not reduce all value to market prices and all human activity to commerce. Ecologically, this approach involves the recognition of the value of diversity in itself. All life forms have an inherent right to life; that should be the overriding reason for preventing species' extinction."[26]

She goes on to argue that the monocultural approach is inherently destructive of both biodiversity in general, and biodiverse forms of agriculture.

So what do economists and government advisers mean when they talk about agricultural productivity? Sir Donald Curry was asked to provide a comprehensive review of the state of British agriculture and suggest changes to the industry, following the foot and mouth outbreak of 2001. In the UK Government's response to his comments on agricultural productivity, it says: "The UK's slow productivity growth partly reflects a slower pace of restructuring in UK farming. Reductions in the number of farms and in the numbers of people working in farming have generally been at a slower pace in the UK than in the rest of the EU."[27] This is just dinosaur thinking, once again. Continuing to increase the size of farms, continuing to reduce labour and continuing to increase output per hour, has no future as a rational and sane strategy. If one extrapolates these trends *ad infinitum*, one ends up with one farm run by one farmer and several million robots run by a giant central computer, using huge amounts of energy per tonne of food produced, with a poisoned environment, and denuded and sick soils. It is the same outmoded thinking that assumes that the economy should continue to grow for ever! The economist Robert Repetto has noted:

> "Under the current system of national accounting, a country could exhaust its mineral resources, cut down its forests, erode its soils, pollute its aquifers, and hunt its wildlife and fisheries to extinction, but measured income would not be affected as these assets disappeared. . . . The result can be illusory gains in income and permanent losses in wealth."[28]

Nature does not function by these rules. Everything in Nature is governed by the rule of the optimum, the line of least resistance and the recycling of materials and energy—the optimum amount of energy to achieve an aim and no more. Fritjof Capra has said:

> "The nonlinear interconnectedness of living systems immediately suggests two important rules for the management of social and economic systems. First there is an optimal size for every structure, organization and institu-

The Challenges Ahead

tion, and maximizing any single variable—profit, efficiency, or GNP, for example—will inevitably destroy the larger system."[29]

In other words, this approach to running any system has its own destruction built in. Applying these new, more inclusive, ways of judging efficiency and productivity to farming indicates the optimum size of a farm to suit the area and circumstances, sometimes smaller, sometimes bigger. Farms will have to be as self-sufficient as possible in the creation of plant nutrients and pest control, in other words cutting down the farm's imports, rather than having to continuously increase profits per man-hour and hectare so as to pay for the ever-increasing costs of the fuel, fertilizers, pesticides etc. that are necessary in order to increase the yields to pay for the imports, and so on. The results of this vicious cycle is that margins are progressively cut to the bone with inevitable results. As Jules Pretty put it:

> "Food commodity prices have been falling steadily over the past twenty years, and most industrialized countries have moved well away from the threat of food shortages. . . . However, in this success lies the seed of destruction. . . . There are many signs that our highly productive and modernized systems are now in crisis."[30]

By removing subsidies on agricultural end-products and using them instead to reward farmers in direct proportion to how sustainable their practices are, would then be a recognition that ecological and social parameters have to be included when judging whether agricultural systems are truly economic and productive.

Training

One of the major crises for the agricultural industry as a whole is that not enough young people are coming into it. Farming has such an uncertain future at the moment that youngsters are choosing other careers. This is very worrying, because if it continues it will spell the eventual demise of farming. Over the two years from 1999 to 2001 alone, 51,000 jobs were lost from farming in the UK.[31] Organic agricultural courses are becoming increasingly popular amongst the young, however, so let us hope that this will help to redress the balance. With older farmers retiring and dying, they can no longer pass on their skills, especially with so many farmers' children deciding on other careers.

As the organic sector continues to grow and flourish, there is a real opportunity to draw young people into farming via the organic route. The

National Farmers' Union is developing proposals for an advice scheme for new entrants to farming, and the National Federation of Young Farmers' Clubs is working on mechanisms for bringing together new entrants and retiring farmers. These are both positive moves, but there needs to be more recognition by the government of the seriousness of the problem.[32] A growing number of university and agricultural colleges in the UK now run courses in organic farming; Aberystwyth Agricultural College, University of Wales, for example, has a history of focusing on sustainable farming and ecological studies. Training programmes are also increasingly being established in many of the EU countries, notably Denmark and Italy, often with government support.

Comparative Yields

I recently heard a supermarket adviser say that if all crops were grown organically there would be a 50 per cent drop in yield. Where did he get this figure? This is the sort of figure which is now treated as an accepted and unquestioned 'fact' not only by those who question the advocacy of organic farming, but also by journalists. One needs to stop and say: "Who says so? What evidence is there for this assumption? On what parameters are we basing these assumptions? How valid were these experiments?" and so on. There is no genuine research that I have read to back up this figure. As Nicolas Lampkin has said:

> "In organic farming systems, crop yields are frequently higher than would be expected if the conventional farming textbooks were to be believed."[33]

Many research studies have been done in different parts of the world comparing the yields of organic and sustainable farming methods with modern conventional farming, and the comparative figures are very variable. Here we are dealing with Western industrial societies, with similar forms of agriculture and less variable climates, but even so there are still differences in the figures. The Soil Association's official figure is a 20 per cent reduction in organic production compared with conventional farming yields for the UK. One thing is for sure, however: the truth is not as clear-cut as one would like it to be, but I will endeavour to sort out the wood from the trees.

The first question to ask about any comparative research is: "How long had the organic farms in the survey been organic?" This is critical, because for the first few years the yields drop while genuine soil fertility is being restored. Research in Germany, published in 1990, showed that yields on farms converted to organic production increased fastest over the first few years, but continued to increase over seventeen years before starting to level off.[34] If every farmer went organic overnight, which is not likely, there would be a noticeable, but temporary, drop in yield. When the conversion period is over and the new system is bedded in, then much more acceptable yields result.

The second question to ask is: "Which crops are you comparing?" Conventionally grown wheat will almost always outstrip organic wheat. There have been dramatic increases in yields of conventionally grown wheat in recent years and although organic yields have continued to improve, they have not kept pace and probably never will. Differences are often around 20–25 per cent, and it is possible that the contention that organic yields are generally 20 per cent less than those on conventional farms comes from this single comparison.

According to Nicolas Lampkin, when comparing the growing of grain legumes (beans etc.) then there is little or no difference in yields. With crops like carrots, beetroot and many other types of horticultural crops, the yields of organically grown ones are at least the same—often better. Again, with milk production per cow, the yields are often similar, or slightly less; however, in this case the production of milk per hectare is around 23 per cent less because organic farms are less intensive.[35]

Typically, organic farms tend to be mixed farms, and even organic dairy farms will include other animals such as lambs and pigs, as well as the growing of specialist cash crops such as milling wheat, porridge oats etc. As a result, comparing yields between farms which are not always of comparable size and type is misleading: for instance, the comparison of yields of wheat grown on a mixed organic dairy farm and an intensive eastern county grain farm. In 1975 a truly comparable study was done by the Cambridge University Agricultural Economics Unit, on behalf of the Soil Association. It studied six organic farms between 1973 and 1974. All were mixed stock and arable farms with dairy herds. Cereal and milk production were the main factors studied, and they were compared with similar conventional farming enterprises. The yields of wheat, barley and oats were all found to be very similar.[36]

One thing I have noticed over the years when studying any comparative research is that where the organic farms involved use biodynamic methods, the yields tend to be significantly higher than those of other organic farms and are consequently more likely to be comparable to conventional yields.[37]

One further question needs to be asked when discussing comparative yields: "When comparing yields, shouldn't we take other factors into account, such as the cost and energy expended on imported fertilizers, the cost of cleaning up polluted water and the environment, and the destruction of soil fertility—in other words, the cost-effectiveness of the different methods?" Some would still continue to argue that we cannot afford to lose 10–20 per cent in yields; the answer to that is that continuing with the unsustainable forms of farming that are prevalent today is truly unaffordable.

Moreover, EU farmers are paid around £2.4 billion per annum to leave land out of use because of overproduction, due directly to intensive forms of farming. If food was grown organically, and therefore extensively, then much of the shortfall in organic yields would be cancelled out.

Subsidies and Tariffs

Many who support organic and sustainable farming believe in subsidies to support organic and more ecological forms of farming. But subsidies on agricultural products, whether crops or animals, and indirect subsidies—such as tax-free farm fuel, financial support for the cost of fertilizers, or the EU's new structure of subsidies based on the area of land farmed—are damaging in many ways. They damage both Western and Third World farmers at the same time. They create unrealistic, 'Mickey Mouse' economics, which has completely distorted agricultural production.

The inevitable result of these subsidies has been to produce huge surpluses of certain food items, grown in unsustainable ways, which are then dumped on the world market at unrealistic prices, undercutting Third World farmers' produce and damaging their livelihoods. Tariffs on certain Third World food products then exacerbate the problem. For instance, US cotton farmers are subsidized to the tune of $4 billion dollars per year for crops that are only worth $3 billion. This has the effect of reducing the world price of cotton by 25–26 per cent, and causing serious damage to cotton farmers' incomes in other parts of the world. Just cutting the US subsidies would make cotton farming profitable again for many in the Third World by hundreds of millions of dollars per year. Another example of this unfair support for agriculture has been the European Union's subsidies for their sugar farmers. EU sugar beet producers have been getting massive subsidies, amounting to 2–3 times the world market price. As a result, there have been surpluses of European sugar. The EU has then subsidised the export of these surpluses, resulting in a further depression of the world price of sugar. A further result of this economic insanity is that EU consumers have been paying an extra £8 per week on their food bills, to pay for the support of EU sugar. Fortunately as part of the reforms to EU farm subsidies, the guaranteed price on sugar will be reduced by 39 per cent over two years from 2007.

Both the EU's and the USA's agricultural policies massively support inorganic, conventional agricultural production, either through direct or indirect methods. They encourage over-production, through chemically grown crops and intensive animal husbandry.

Fortunately there are moves in the right direction, in the EU at least. From January 2005, subsidies on end-products are being phased out to be replaced by a flat rate per hectare as long as farmers follow certain limited procedures, such as protection from soil erosion, the maintenance of organic matter and soil structure and other environmental and landscape practices. The more environmental practices they follow, the more money they will get. Failure to comply will result in possible reductions in the payments.[38] However, there has been a major compromise, due in large part to the power of the farming lobby. Although subsidies on end-products will be abolished, this has been replaced by a subsidy on agricultural land of £50 per acre. A preferable model, in my view, is the one adopted by the Swiss government, where comprehensive and graded support is given to Swiss farmers. They have abolished all forms of general subsidies, replacing them with incremental rewards only to those farmers who introduce more ecological forms of farming enterprises, with the highest support of all given to organic farmers.[38]

One has only to look at the level of financial support for agriculture amongst the thirty richest nations who are members of the Organisation for Economic Co-operation and Development (OECD) to see that these richest nations remain intent to support their agricultural industries at the expense of the poorer countries of the world who cannot afford to do so. In 2004 the total support of all OECD members was a staggering $378 billion. However, there are the examples of OECD members, such as Australia and New Zealand, where subsidies have been all but eliminated, which has proved that the end of subsidies has not brought doom and gloom and not destroyed their agricultural sector. Indeed, after the initial shock and readjustment, both countries are doing well, having created new markets in the Pacific Rim.

What is being proposed here is the abolition of subsidies on the end-product, on fertilizers and fuel, which distort the market and encourage destructive forms of overproduction; whilst retaining subsidies that aim to reward more ecological approaches, to help with diversification schemes, schemes for renewable energy production, or financial support for the conversion to organic farming and other sustainable practices.

Co-operatives

Co-operatives and other similar ventures are just the kind of developments that can provide genuine improvements in productivity and economies of scale, without destroying the family-farm structure and optimal farm size. The UK Government's Strategy for Sustainable Farming and Food strongly suggests that co-operatives are a good idea, commenting that "the UK lags behind its counterparts in the EU in both the number and scale of co-operative ventures." In the dairy sector we are level with other countries, but in every other sector we lag behind, especially when it comes to farm inputs, where good savings can be made by buying in bulk.

Areas of France like Normandy and Brittany have very successful co-operatives. The farmers have recognized that what they are good at is growing quality food. What they are not good at, and indeed do not have the time to do effectively, is marketing their goods. So they band together and the co-operative pays a professional manager, or a team with proven marketing skills, to run it. The manager finds markets and obtains contracts for the following season; the farmers then decide amongst themselves how they are going to supply those contracts. They know that the products they will produce the following year already have a market, and they know how much they will be paid. Many of the ventures have cheese-making factories and meat-processing plants, which manufacture high-quality local produce using local recipes. They use their purchasing power when buying in materials, fuel and equipment to keep the prices down, and when negotiating with the supermarkets, because the latter know they will get regular supplies of a certain quality. This system provides better security for the farmers, and improved consistency of supply and quality of local produce for the supermarkets and benefits to the customers.

Permaculture

Like zero-tillage, permaculture uses the minimal amounts of energy and as a result points the way in a fuel-short future. The concept is a modern one—or rather a modern reincarnation—developed in Australia in the 1970s by two ecologists, Dr Bill Mollison and David Holmgren, and first described in their book *Permaculture One* in 1978. It is now well-established across the world, and over the years has proved to be applicable to every climatic and cultural zone. It essentially involves growing as many perennial crops as possible: well-spaced trees with an understorey of shrubs, both providing crops of some sort, whether fruit, nuts, seeds, useful timber products or fuel. Areas are set aside for both pasture and cropping areas, using zero-tillage techniques, as well as ponds for fish production. The whole aim is to make the farm as self-sufficient as possible using the least amount of energy.[40]

As with most human activities, the idea is not new. One notable example of permaculture in action occurred in the forested area around Mount Kilimanjaro in Kenya, before the white settlers arrived. The people had developed a three-dimensional, highly productive form of sustainable permaculture, which consisted of well-spaced-out trees and shrubs, carefully selected over generations, that were useful for both food and medicine, and with ground crops in the more open areas. Their huts were built up in the trees, which allowed for a high population without limiting the land available for growing food. This system supported a very heavy population for hundreds if not thousands of years.

Another example, which still exists, is the medieval system called *montados* and *dehesas* in Portugal and Spain. It is characterized by species-rich permanent grassland, thinly planted with mostly cork and holm oak trees, under which are grazed sheep, pigs and cattle. The pigs and sheep eat the acorns, and the cork trees are harvested for their cork. In some areas the wider avenues between the trees are temporarily cultivated in rotation to grow grain, such as wheat or barley, and other crops to help feed the animals.[41] See also the section on forest farming in this chapter.

Organic Farming Without Animals

All of the modern organic pioneers of the last century—and indeed many of its advocates today—believed that a vital part of any organic enterprise must be animals. When Sir Albert Howard looked at any natural system like a forest, he saw that "mixed farming is the rule: plants are always found with animals: many species of plants and animals all live together."[42] The accumulation of humus on the forest floor was the result of both organic plant materials and small percentages of animal dung. He was convinced that for the production of the best humus and plant food, there had to be animal manure, but even more important was urine.

"The key substance in the manufacture of humus from vegetable wastes is urine. . . . It contains in a soluble and balanced form all the nitrogen and minerals, and in all probability the accessory growth substances as well, needed for the work of the fungi and bacteria which break down the various forms of cellulose—the first step in the synthesis of humus.[43]

Rudolf Steiner also believed that animals, especially cattle, were an essential part of the make-up of every farm. However, many would argue today that to feed the world organically and sustainably a much higher proportion of cereal, beans, vegetables and fruit must be grown than is often suggested by organic advocates.

Although it might be argued that to produce food this way would diminish the vitality of the soil, I am not convinced. There has to be a place for low-stock and even animal-free organic husbandry if we are going to take seriously the task of supplying all the food required for the world's population.

There are numerous examples of successful organic production with few or no animals. In Guatemala and Honduras, a combination of zero-tillage, the growing of Mucuna beans to build fertility and the return of all crop residues to the soil has been extraordinarily successful in improving soil fertility and crop yields. Then there are the age-old traditional methods used by small-scale women farmers in areas like the Indian Punjab, who have produced food intensively over many centuries with little or no animal manure. Moreover, if we study the traditional farming of the East, as described in King's book *Farmers for Forty Centuries*, we find that they managed to grow food organically and intensively over several thousand years with a minimum of animal and human manure.

A modern example is that of rice farmers in many parts of Asia who grow the small fern azola that floats on the surface of the paddy fields and which has nitrogen-fixing algae that grow on the leaves, keeping down the weeds and providing compost and nitrogen to grow the rice. There are around 36,000 ha (89,000 acres) of rice fields using this system in the Fujian province of China alone. Other areas of Asia where this technique is used are the Philippines, Vietnam, India and Thailand. In some areas fish are added to the paddy fields, helping to provide manure, oxygenating the water by their activities and providing a good source of protein.

In the UK, there is the fine example of Iain Tolhurst, a vegan who eats no animal products of any kind, who has managed to build up the fertility on what was fairly poor ground near Reading. Here he produces abundant crops, distributing 300 vegetables boxes each week to Reading and Oxford. He grows different brassicas, leeks, onions, leaf beet, Swiss chard, beetroot, celery, celeriac, carrots, parsnips, a whole range of legumes, courgettes, sweetcorn, squashes and, of course, potatoes. He manages this by putting in a red clover green manure crop one year in nine, and regularly growing overwintered and undersown green manure crops of clovers or cereals throughout the rotation, as well as making plenty of compost without using manure.[44]

There is also a modern version of the highly intensive traditional small-scale farming of India: 'biointensive' gardening and mini-farming as advocated by John Jeavons in his book *How to Grow More Vegetables, Fruits, Nuts, Berries, Grains and Other Crops Than You Ever Thought Possible on Less Land Than You Can Imagine*. His ideas have been adopted around the world, as they are applicable to both the West and the Third World. There are five components:

1. *Double-dug, raised beds.* The soil is first trenched 30 cm (1 ft), then loosened a further 30 cm. This incorporates air into the top 60 cm (2 ft), allowing plant roots to penetrate easily. The raised beds are 1–1.5 metres wide.

2. *Intensive planting.* Seeds or seedlings are planted in the raised beds using hexagonal spacing, in such a way that their leaves will touch when grown, providing ground cover. This results in a 'mini-climate' under the leaves, which retains moisture, protects soil microbiotic life, keeps weeds down and produces high yields.

3. *Composting.* Compost is used, often increased by growing crops specifically for the purpose.

4. *Companion planting.* Maize and beans may be grown together, for example green beans and strawberries. Other plants are mixed in which attract beneficial insects or have a repellent effect on certain pests. Some wild plants are allowed to grow which have deep roots to 'mine' the subsoil, loosening it and bringing up trace elements and washed-out nutriments. In other words, as much biodiversity as possible is created for the maximum health of the system.

5. *Whole-system synergy.* Biointensive food production only works when all the elements described above are carried out together.

On a larger scale, cereals and other grain crops could be grown using a combination of the methods used in Guatemala and Honduras—legume green manures and zero-tillage—as well as sheep in an adapted form of the old Norfolk four-course system. There are, I believe, many imaginative ways that organic and sustainable farmers could grow crops with few or no animals while at the same time increasing soil fertility, whatever the doubters may say. This is not only desirable but imperative if we are to feed the world sustainably.

Forestry

Not only will coppice systems for energy production have to become more common, but it is essential to grow as many trees as possible for timber production as well, as we are currently cutting down more trees than we are planting. Many companies claim to be producing timber sustainably by planting two or three trees for every one they cut down. But what they are doing is destroying trees weighing many tonnes each and replacing them with saplings that will take fifty years or longer to grow to maturity. There are sustainable forestry systems that take this time lag into account, making sure that the loss of biomass through felling occurs at the same rate as the increase in biomass through new growth.[45]

On a smaller scale, mixed forestry can be profitable if it is coppiced and close-planted initially, and a value-added scheme is used. So the original planting would be of hardwood trees of different growth rates, infilled with coppice trees that will be cut on a regular basis once established, as well as short-term fast-growing conifers to provide quick returns. The coppice trees would be the first to come into production, then the conifers, followed by the faster-growing hardwoods. Then the longer-term hardwoods would come into their own. As the hardwoods are cropped, they are replanted. If there is a factory on site, a wide range of products might be manufactured, from small objects made from the thinnings and coppice-wood, hurdles and baskets from willow whips and waste wood, and larger turned pieces, garden furniture and high grade pine and hardwood furniture from the main timber. In this way maximum income is derived from the woodland in the shortest possible time. For farmers who have land that is not suitable for more intensive forms of agricultural production, this approach provides an additional or alternative income.

One of the recommendations of the Strategy for Sustainable Farming and Food is for the Forestry Commission's Woodland Grants Scheme to fund the creation of new woodlands and improved management and regeneration of existing woodlands. Woodland creation is also stimulated through the Farm Woodland Premium Scheme and the work of the National Forest Company and the Community Forests.[46]

The Soil Association's Recommendations

To summarize, I cannot do better than quote from the Soil Association's response to the UK Government's Policy Commission on the Future of Farming and Food 2001. It outlines an interdependent strategy to revitalize and transform agriculture in the UK, but it could apply to any Western country. It lays down strict criteria which will be difficult to achieve, but nonetheless provides an essential set of goals to aim for if we are to have truly sustainable agriculture in the future.

Environment
- Farming systems which are genuinely sustainable, in terms of the use of finite natural resources and their impact on the environment
- Britain's farmland wildlife to be restored to a diverse and abundant state
- No pollution of watercourses and aquatic ecosystems
- Healthy soils, fertile, with good structure, minimizing erosion and reducing water run-off and thus flooding downstream
- Agriculture and the food industry's contribution to the acidification of the soil, water and atmosphere to be minimized
- Agriculture and the food industry's contribution to global warming to be minimized

Health
- Food free of pesticide and veterinary drug residues
- Less sugar, salt and other undesirable additives such as hydrogenated fats added during the processing of food
- No routine farm use of antibiotics, minimizing the development of antibiotic-resistant bacteria
- A reverse in the decline in nutrient content and taste of fruit and vegetables
- A substantial reduction in incidents of food allergies, intolerances and poisoning
- Confidence that all human degenerative and other diseases that may be linked to food quality are being avoided, whether links proven or not

- Genuine application of the precautionary principle which justifies preventative measures in the face of scientific uncertainty of the level or nature of risk
- More emphasis on developing agricultural systems that produce healthy, safe and nutritious food than on further 'hygiene' methods
- Food production to be based on natural processes
- No use of genetic engineering in food production
- An end to major national food health crises caused by industrial practices

Animal welfare
- High standards of animal welfare
- No animals kept in confined conditions
- Minimal long-distance transport of live animals

Farming communities
- Farming to be a prosperous and well regarded sector
- A reversal in the fall in farm employment
- A continued and central place for small and family farms

The economy
- Public funds to be used only for achieving public benefits
- The total cost of food production (direct subsidies plus direct costs) to the taxpayer to be minimized
- Regeneration of rural and local economies

The food chain
- Farming methods which use the fewest agro-chemicals and other pollutants, to be most encouraged
- A diversity of crops and livestock, including traditional breeds, rather than monocultures
- A major reduction in food miles; a generally localized food chain

Traceability and accountability
- The majority of food that can be, to be produced in season in the UK
- International trading systems which avoid exploitative practices in developing countries and reduce hunger, malnutrition and poverty
- Confidence and assurance that farmers, the food industry and Government have the interests of the public and environment at heart.

The Future for Third World Agriculture

Traditional Farming

As Jules Pretty has described, there are two types of agriculture in Third World countries: traditional, often sustainable systems, and 'Green Revolution' systems. The traditional systems are "usually located in drylands, wetlands, uplands, savannahs, swamps, near deserts, mountains and hills, and forests." [47]

Crop yields in traditional farming

Who farms and maintains these farms, this 'forgotten' agriculture which supports some 30–35 per cent of the world's population? As Vandana Shiva has said, it is small-scale farmers, most of whom are women. For this reason they have largely been overlooked, or more accurately are completely invisible to the largely male agricultural strategists in organizations like the UN, the World Bank, and the World Trade Organization.[48] Yields are usually seen in terms of monocrops grown on large areas, and as we have seen, when measured in this way conventional farming appears to be more productive, but when the yields of all the crops grown together in a traditional mixed system are taken into account, they often outperform even the most productive orthodox monocrop. It was this ignorance on the part of agricultural specialists and advisers that Vandana Shiva described as "this blindness to the high productivity of diversity, a 'monoculture of the mind' which creates monocultures in our fields and in our world."[49]

She went on to explain how in Java, small-scale farmers cultivate 607 species in their home gardens; in sub-Saharan Africa, women cultivate 120 different plants; and a single home garden in Thailand typically has 230 species. She mentioned a study in eastern Nigeria which found that the home garden, usually cultivated by the farmer's wife and occupying only 2 per cent of the farmland, accounted for half of the farm's total output; and how in Indonesia 20 per cent of household incomes and 40 per cent of domestic food supplies come from home gardens managed by women. She explained how research done by the Food and Agriculture Organization (FAO) of the United Nations has shown that small biodiverse farms can produce many times more food than large, industrial monocultures.

Jules Pretty's Studies

Jules Pretty has studied the various attempts over the years by external organizations to impose on Third World farmers Western methods and ideas. His conclusions are that any attempts to help Third World farmers, must recognize existing traditional systems. A large part of his work has been to collate examples of successful traditional farming methods, and cases of local people helping each other to improve or regenerate sustainable systems. One of his objectives has been to make governments and the international community aware of the value of traditional systems developed over the centuries, involving local knowledge of the most suitable crop varieties and of the terrain and local climate, and combining this local knowledge with suggested improvements to these traditional systems.

A fine example of successful traditional methods being recognized and combined with expert help, is the Cordillera area of the Philippines. The rice growers of this area are well-known for their exceptional seed collectors, who have managed to maintain and improve on the local strains of rice over many generations. Before the main harvest, between one and three women walk through the terraced paddy fields, selecting seed only from the healthiest, most robust, and most disease-free plants. Only when enough seed has been collected to sow the following year's crops does the main harvest begin. This form of rigorous selective breeding over countless generations has ensured relatively disease-free crops that are best suited to the region and local conditions.

Experts from the European-funded Central Cordillera Agricultural Programme (Cecap) noted that when some local farmers tried new lowland hybrid varieties, they failed. Nevertheless, they became aware during their studies of the low economic returns and decreasing rice production on these highland farms.

Their conclusion was to combine the best of the indigenous selective breeding techniques with a change in the transplanting time. Farmers in the area usually planted their seedlings out into the paddy fields one to two months after germination in the seed beds. Cecap found that if they transplanted them at the ten-day stage, they produced more shoots and a 20 per cent increase in the final yield. According to Cecap, the local growers have adopted the techniques and are themselves promoting it now among other farmers.[50]

Cecap has also encouraged other 'indigenous best practices', such as using composted sunflower haulms, and as previously mentioned, the use of the crinkly fern called azola, which grows on the surface of the water in the paddy fields, both smothering weeds and providing nitrogen for the rice when harvested and composted or used as a green manure.[51]

Pretty has also made studies of cases where local people, local authorities and even in some cases national institutions have sensitively helped in supporting sustainable systems with co-operation and input from the farmers themselves. Sometimes this involves helping with improvements to already existing systems, sometimes reviving systems that have been abandoned. Often it involves staff working and living with locals, learning from them and informing them about successful projects elsewhere that might be appropriate in their area and circumstances, as well as finding ways of improving existing systems. Subsidies are often provided for suitable capital projects that would otherwise be beyond the locals' means, such as water conservation and irrigation, or the construction of locally designed low-tech equipment suitable for their needs.[52]

Much of his research has involved establishing why many of the previous projects failed, so as not to make the same mistakes again. His conclusions are various. The obvious mistake was the imposition by outside agencies of high-tech solutions involving expensive fertilizers, pesticides and equipment. However, he also found failures where the ideas were sensible and appropriate, but were imposed from above without the co-operation and input from the local farmers.

In summing up, Pretty described how when the process of agricultural development focuses on resource-conserving technologies, local groups and institutions, and enabling external institutions, then sustainable agriculture can bring both environmental and economic benefits for farmers, communities and nations. He went on to suggest, that until recently, most of the empirical evidence was limited to findings from research stations, farm trials and demonstrations, but that the most convincing evidence comes from the successful and practical applications in real situations in countries of Africa, Asia and Latin America, where hundreds, and possibly thousands, of communities have now regenerated their local environments.[53]

Forest Gardens

The forest garden is a tropical form of permaculture which has been successfully applied in Sri Lanka, Ecuador, Indonesia, the Philippines and Kenya.[54] It involves encouraging useful crop species in jungles and forests using organic methods, which can be adopted by small and large farmers alike. Experience shows that this form of sound ecological land management improves the economic conditions of rural communities. Where it has been instigated it has brought about agricultural sustainability, biodiversity, environmental protection and stability and economic benefits. It involves growing several commercial crops at a time, thus spreading the risk for the farmers.

In Sri Lanka, the Neo-Synthesis Research Centre (NSRC) have been instrumental in carrying out experiments in analogue forestry. The aim was to help stop the destruction of tropical rainforest by replanting the forest where it has been felled, using mainly trees and shrubs that yield valuable products. All the crops are grown using organic methods. The result has been a diverse forest environment, very like its natural counterpart—so much so that many species of animal and bird have moved in from the surrounding forest to set up home. This kind of sustainable farming is helped by the growing market for organic products in industrial countries, especially those products that cannot be produced in temperate regions.

Fair Trade and Organic Exports

It should be emphasized that not all fair-trade products are organic. However, where they are, they offer the best of both worlds. As Lynda Brown explained in *The Shopper's Guide to Organic Food*:

> "Fair-traded organic teas and coffees offer the best of both worlds: as well as being organic, they are guaranteed to have been produced under conditions that offer maximum benefits to the employees and their communities—above-average wages, training, protective clothing, housing, a clean water supply, good living conditions, health and safety, profit-sharing and a pension scheme. Companies who buy fair-traded products pay above the market price, part of which goes into a social fund for the workers and their community." [55]

The increasing popularity of organic tea is transforming plantations like the Uva Estate in Sri Lanka's highlands, which has been organic since 1995. It was originally owned by the Government and used conventional methods. There were problems with soil erosion and chemical pollution, and a marked reduction in wildlife. Since the introduction of organic methods the number of birds, reptiles and amphibians has risen dramatically, which helps to keep pests at bay. In fact there are very few problems with diseases and pests now that the soil is back in good heart and there is increased biodiversity on the estate. This kind of project is helping to improve the lives of many

Third World farming communities as well as their local ecosystems.[56]

In July 2001, the EU's new banana import regime came into force, owing to pressure from the US government through the World Trade Organization. Previously the EU gave preferential treatment to former colonies which relied on bananas for most of their external income, rather than Latin America, where US multinational companies have a large financial stake. As a result, the livelihoods of the small farmers of the Caribbean particularly in St Vincent, Dominica and St Lucia, are threatened. There is now a move to encourage them to become organic. A number of farms have been certified by a UK body and are now producing organic bananas and other produce. One of the advantages that these islands have is that they are free of black sigatoka, a serious disease which is threatening world banana production.[57]

Turning the Tanker Around

Changing the direction of agricultural production to make it truly sustainable will be a Herculean task, but as many examples show, it can be done. Whether it is European governments such as Switzerland deciding to take the ecological route, or consumer demand in the UK, whether the changes are coming from small peasant farmers in South America or India, or whether the changes are being brought about by government policy as in south and south-east Asia and China does not matter. What is certain is that the momentum is growing, as more and more farmers, governments and members of the public begin to see the benefits of more sustainable practices. For the industrialized countries and 'Green Revolution' areas of the Third World, it is going to be more difficult, but the traditional small farmers, who feed a third of the world's population, already have a head start. All they need is either to be left to get on with what has often been very successful food production over many generations, or to be given sympathetic help and support to improve what are often already sustainable traditional practices.

Chapter 9

The Organic Uplands: A Vision of the Future

"Earth is a Goddess
and teaches justice to those who can learn,
for the better she is served,
the more good things she gives in return."
—Xenophon

The Dinosaurs and the Mammals

Around 70 million years ago, towards the end of the Cretaceous period, dinosaurs filled practically every ecological niche on Earth. However, a new race of small animals called mammals was quietly evolving in their midst—and would soon replace the apparently indestructible reptiles. Could a similar phenomenon be in progress in agriculture?

Two Farming Systems
In the southern states of America, an endless sea of maize stretches out in all directions to the horizon. This is the hungriest crop there is, consuming more nitrogen fertilizer and chemical herbicide than any other crop. The pollution from its production, along with other crops, finishes up in the Gulf of Mexico, where it has poisoned a huge area of the sea, creating a dead zone 31,000 sq km (12,000 sq miles) in extent. Much of the harvested maize has one purpose, and one purpose only: to produce beef.

Meanwhile, in the middle of the Kansas prairie, for as far as the eye can see there are endless steel-fenced cattle pens, each containing 150 animals, standing or lying listlessly on their own excrement. There is no grass to be seen. There is nowhere for the cattle to run, nowhere to go, nothing to do except eat, rest or stand. The endless pens are criss-crossed by dirt roads. The stench is all-pervasive. This 'farm'—better described as an outdoor beef production line—houses approximately 75,000 head of cattle—400 per hectare (162 per acre). Each collection of pens has its own vast waste lagoon in the middle to accommodate the 75,000 tonnes of excreta produced every year. At the very heart of this outdoor factory stands a huge industrial feed-mill. For the first six months of their lives, the calves are with their mothers on grass. After a traumatic weaning they are taught to eat maize exclusively—not cattle's natural food. Because of this unnatural diet, a powerful antibiotic is mixed with their food to stop them getting diarrhoea. This will continue every day for the rest of their short lives—fourteen to sixteen months. Growth hormones and protein supplements complete their diets.

These 'farms' use huge amounts of energy to produce their meat. It has been estimated that producing the fertilizers to grow the corn, the milling and all the other energy needed for one beef animal takes 1,292 litres (284 gallons) of oil. The final irony is that the meat is sold as 'ranch beef'![1]

In the beautiful Cotswold Hills, in Worcestershire, is Kite's Nest, a mixed organic family farm, surrounded by intensive cereal farms which fifty years ago were similar. The main product is beef. The pastures are many generations old and full of wild herbs and rare flowers, such as pyramid orchids, clustered bellflower and dyer's greenweed. All the cows have names and live in family groups feeding on the rich mix of grasses, clovers and herbs, the last of which they seek out with relish. The calves are not forcibly removed from their mothers; there is no need to. When its mother's milk dries up in late pregnancy, the calf is weaned naturally. The cows live longer than they would do under more intensive conditions, and with their reasonably stress-free lives and organic diet, they are healthier. When the cattle are old enough to slaughter they are taken to the local abattoir, and the meat is sold in their on-site farm shop at premium prices.[2]

The 'Green Revolution' and its Alternative

In the 1960s many North African countries, the Punjab and Haryana in India, parts of Latin America, and parts of southern, south-eastern and eastern Asia were gripped by a new fervour. The 'Green Revolution' was a visionary plan by the West, the UN and many Third World governments to 'modernize' agriculture, partly in response to the regular famines that plagued countries like India at the time. It would employ all the latest scientific agricultural methods, which had proven so effective in the West in vastly increasing yields, especially since the end of the war in 1945. It was simplicity itself, a 'one plan fits all' scheme:

1. Agricultural scientists bred new varieties of staple cereals that had three major advantages: they were shorter-strawed, they were quick-maturing and they responded well to high applications of nitrogen fertilizer. With shorter straw, a higher proportion of nutrients was utilized in producing grain rather than straw. It also meant that when higher levels of nitrogen fertilizer were used, the crop did not grow too high on weak stalks which were then liable to break in the wind and rain. Quick-maturing qualities meant that two or even three crops could be grown in one season.

2. Large-scale monocultures were the norm—rice, wheat, cotton, sugar cane, pineapples, bananas and intensive livestock enterprises and ranching.

3. Credit schemes were set up to enable the farmers to borrow money to purchase the inorganic fertilizers, pesticides, and machinery, as well as for irrigation schemes.[3]

To begin with, yields doubled on average over the years, but the system tied the farmers into spending a large proportion of their income on the new external inputs, not to mention the interest on their loans. Previously there were few external imports. The seeds of the crop varieties, bred over generations to suit local conditions, were saved from previous crops or swapped with neighbours. Manure and other organic wastes from the farm itself provided the nutritional requirements. Local knowledge provided the farmer with systems for pest and disease control, for example the neem tree, which is a natural pesticide.

All those countries where the 'Green Revolution' has had a significant impact have seen a stagnation and in many cases a falling off of crop yields since the initial high levels of the first few years (1965–1980).[4] People with a knowledge of soil ecology could have predicted that adding mineral fertilizers to soils with even reasonably good levels of organic matter and a rich soil life, would increase yields, in some cases dramatically, in the first few years of fertilizer use. At first the complex soil ecology remained intact, but the added chemical nutrients caused the natural stores of unavailable organic nutrients to be released rapidly by increased bacterial activity and nutrient turnover, which burned up the natural resources. Finally the natural system started to degrade and the soil ecology became much more limited, with most of the plant nutrients being supplied by the fertilizers alone.

The same happened with pesticides. To begin with the pesticides had a miraculous effect, but the diseases and pests started to become tolerant and the natural predators were killed off, which reduced the effectiveness of the pesticides and in many cases actually increased the level of attacks. Insecticides, and particularly fungicides, also took their toll on soil life. Combined with monocropping, these factors caused an equilibrium where the soil was more or less biologically dead, or at the very least greatly degraded. The accumulated chemical toxicity and micro-nutrient depletion also played a part. In order to reverse these declines, fertilizer applications have had to be increased by around 50 per cent.

The same thing happened to yields in the first years after the introduction of fertilizers in the West, with a subsequent slight fall-off to more consistent levels as long as fertilizer application remained the same. The reason much of the West has been able to continue to grow crops conventionally over the years, is largely the forgiving soil types and weather patterns we have. However, in many Third World countries the story has been very different.

The Punjab and Haryana were some of the first areas to adopt the 'Green Revolution'. These two states alone provide 80 per cent of India's food. However, this intensive agricultural project is increasingly wreaking havoc. Crop yields have declined alarmingly over the last few years. Many irrigation schemes have literally started to dry up, leading to some areas becoming almost barren. As a result of the loan schemes, farmers are heavily in debt, which they are increasingly unable to pay off. Rural unemployment is increasing and there is a general collapse in social structures and farm economies. Debt in rural Punjab is now running at some 50 million rupees (£630,000) and on a more personal level, 250 farmers have committed suicide in a two-year period.[5]

Unsuitable crops like rice were grown in what are essentially arid areas, made possible by the new wells that were sunk all over the area. The water table has dropped in many places below the bottom of the boreholes and quite often the water is now saline and unusable. The heavy use of fertilizer has depleted the natural fertility of the soil and the plants not only absorb the extra nitrogen, potash and phosphate but use up the natural reserves of micro-nutrients, which are essential for both soil and crop health. The diminution of micro-nutrients has also greatly reduced the plants' ability to absorb plant foods. Much of the ground water is polluted by excessive nitrates and pesticides, making it increasingly dangerous for human consumption.[6]

One of the results of this drive to industrialize farming in India, as in many other countries, is the reduction in the number of crop varieties. It is estimated that a hundred years ago more than 30,000 varieties of rice were cultivated in India alone, all with unique characteristics which suited the local conditions. This valuable extensive gene pool has been reduced dramatically over the years, and at the present time a mere ten varieties provide 75 per cent of the annual harvest.[7]

Despite all this, the Indian government's response is to push for even more intensive agriculture and to approve the commercial growing of Bt GM cotton and large-scale agri-business projects growing sugar cane, all of which will probably have even more devastating effects on the region.[8]

Jatan

Disillusionment with the 'Green Revolution' has led to a modern organic movement in India. Initially—and very importantly—it was a farmer-led movement, but it has evolved to a stage where universities and even the Government have become involved. Kapil Shah is one of its guiding lights. He was trained in conventional agriculture but has also knowledge of traditional methods, and founded Jatan, meaning 'care', one of India's leading organic organizations. Its main purpose is to link farmers, consumers, universities and government bodies to promote and coordinate organic research and practice. Its formative principle is that of *sajiv kheti*, a natural farming system based on Gandhi's philosophy of simplicity, respect, and a high regard for manual skills. The ten main concepts are:

1. Faith that Mother Earth has enough resources to fulfil the basic requirements of all living beings

2. The belief that nature is the best farmer

3. The use of local resources at the best possible level

4. Mixed cropping and maximum biodiversity on the farm

5. Efficient management and optimum utilization of natural resources

6. Closed cycles of nutrients and other resources

7. Need-based crop selection

8. Minimal reliance on market forces
9. No use of synthetic chemicals
10. Efficient and skilful farm management

Many of these principles are similar to Sir Albert Howard's ideas and the basic precepts of IFOAM as laid out in Chapter 1. So although organic movements in different countries often stem from different roots and have their own local history, the basic principles are immediately recognizable.

Jatan runs training courses, publishes a quarterly newsletter, and organized the second Gujarati meeting on organic farming, which was attended by 800 people, including thirty small and poor farmers, who paid their fees in kind, namely 5 kg of grain that they themselves had produced. Kapil Shah is also Gujarat's representative on Agricultural Renewal in India for Sustainable Environment (ARISE), a national network of organic farmers, environmentalists, voluntary workers and scientists. Vandana Shiva, a vociferous campaigner for traditional women farmers and sustainable agriculture, is one of its co-founders. Gujarat is also one of the major centres for organic cotton production in India, the products of which are becoming increasingly popular around the world. Growing organic and pesticide-free cotton is also becoming more popular with farmers themselves, with research confirming what many of them already know: that it is more profitable for the farmers to grow organic than conventionally grown cotton.[9]

Positive Developments Around the World [10]

The EU

Whilst conventional farming continues to dominate European agricultural production, organic farming has grown continuously over the last ten years, largely led by increasing consumer demand, although there has been a slight tailing off after the peaks of 2000-1. In the EU as a whole there were 3.3 million hectares of land under organic cultivation in 2000, just under 3 per cent of total farmland—a 20 per cent expansion since April 1999.[10] By the end of 2003 (the latest available figures) this had risen to 5.6 million hectares, with 3.4 per cent of total farm land registered organic, comprising around 143,000 farms.

In 1999 Italy was the fifth largest in terms of area farmed; in 2004 it became the largest. But as a percentage of agricultural land run organically, Liechtenstein tops the bill with 26.4 per cent, followed closely by Austria with 12.9 per cent and Switzerland with 10.27 per cent. Below are the latest available figures for the top twelve countries in both the EU and EFTA.

Country	Organic Area (ha)	%	Organic Farms	%
Austria	328,803	12.90	19,056	9.5
Czech Republic	254,995	5.90	810	2.1
Denmark	165,148	6.14	3,510	6.1
Finland	159,987	7.22	4,983	6.6
France	550,000	1.86	11,377	1.7
Germany	734,027	4.30	16,476	4.0
Hungary	113,816	1.94	1,255	3.2
Italy	1,052,002	6.86	44,043	2.0
Liechtenstein	984	26.40	43	20.5
Sweden	225,776	6.83	3,562	4.2
Switzerland	110,000	10.27	6,445	11.1
UK	695,619	4.42	4,017	1.7

The Danish Government instituted a five-year plan in 1999 to support product and marketing development, training and research through co-operation between government, producers and retailers. Germany remains the largest consumer of organic produce, comprising nearly 30 per cent of European organic sales. Total European sales rose 8 per cent in 2002, and by the end of 2003 had reached 10–11 billion Euros. In the Netherlands, the Ministry of Agriculture launched a four-year strategy whose main thrust was to increase conversion payments, backed up by support for market development, and educational and information programmes for the general public.

Italy, with its government's support, has had the fastest growth rate for organic production until recently. They now have 1,052,002 hectares of land organically farmed, with 44,043 registered farms. The government has introduced an inter-regional programme to support organic production, including training initiatives.

The French government, having realized they were lagging behind in production, whilst at the same time seeing a fast growth in home retail sales, introduced a five-year plan in 1998 to support the development of the organic sector. It included the doubling of the financial assistance for converting farmers, and explicit targets to increase the number of organic farms to 25,000, converting 1 million hectares by 2005. This was somewhat overambitious, but growth has been impressive with a rise from over 200,000 hectares in 1998 to over 550,000 hectares in 2003, and still rising.

The United Kingdom

Consumers have forced the pace in the UK, with demand far outstripping production. As a result, in 2000, 70 per cent of organic food consumed in the UK had to be imported. Research has shown that the market grew on average by 40 per cent per year. Sales of organic food in the year 2000 rose by 55 per cent to £600 million (840 million Euros). In 2002 this had risen to just over £1 billion (1.44 billion Euros). During 2000, 75 per cent of households purchased organic produce, and 30 per cent of all baby foods purchased was organic. Although these rates of growth have eased off somewhat of late, growth in this sector continues.

Questionnaires in 2001 by MORI and NOP found that:

- 85 per cent of the public think that the Government should be doing more to encourage organic food

- 78 per cent think a target should be set

- 65 per cent think more than 10 per cent of UK's farming should be organic

- 52 per cent think it should be more than 30 per cent

- 11 per cent think all farming should be organic.[11]

As the Soil Association's response to the Policy Commission on the Future of Farming and Food said:

"Organic food is popular with the public and this popularity is growing. Modern society's views on a range of issues, from protecting wildlife and natural resources to animal welfare and connections between food and locality, and including attitudes to risk and health, are all moving in a direction which increases support for organic farming and food. All the signs currently are that these trends will continue." [12]

In response to this growing demand, there was almost a doubling of farm land under organic cultivation between 1998 and 2000, from around 270,000 ha to 525,000 ha, and between 2001 and 2002 the area of organic horticultural land in the UK nearly doubled from 2,990 to 5,430 ha. The latest figures (2003) show UK organic farm land at almost 700,000 ha, or 4.42 per cent of agricultural land. It might be worth stating here that organic farms vary greatly in size. Two of the largest organic farms in the UK are Rushall farm in Wiltshire, converted in the 1980s and consisting of 668 ha (1,650 acres), and Woodlands Farm in Lincolnshire, converted in 1996, comprising 680 ha (1,680 acres).

After years of campaigning by the organic movement, in 2000 the UK government at last put forward its Action Plan to Develop Organic Food and Farming in England, as part of its Strategy for Sustainable Farming and Food. The plan recognized the environmental, social and economic benefits of organic farming. It also committed DEFRA for the first time to introducing support payments for farmers after they have converted to organic production. There will also be increased payments for people wishing to grow organic apples, pears and plums. There was a commitment to regionalizing food production, and public bodies will be encouraged to buy organic food wherever possible. Most importantly, the plan aims to increase the market share of UK-produced organic food from 30 to 70 per cent, and commits supermarkets to increasing the amount of UK organically produced food they sell.

The Action Plan for Organic Production in England put forward by the government, states that it is a first stage in a strategy to create a sustainable and competitive organic food and farming sector. The Action Plan:

1. Provides for organic payments to organic farmers to recognize the environmental benefits they deliver
2. Proposes the development of targets for meeting a higher proportion of demand from British sources
3. Improves the market information available for the organic sector
4. Encourages sustainable procurement of food, including organic food
5. Sets aside £5m over five years to support the organic sector's research and development priorities through the LINK programme.

Those farmers wishing to convert to organic farming have, until now, been hampered by the lack of financial support to cover the transition phase. This support is necessary for two reasons. Firstly, for the first two years the farmer may not sell any produce as 'organic', so there is no premium, and secondly, as we have seen, a farm converting to organic production inevitably suffers as the yields drop before the soil recovers its true fertility. Now farmers will be able to claim financial support for this transitional phase, but only after it has been completed. It is much less than the levels of support that our EU partners receive, but it is a start, and it is hoped that it will encourage more farmers to be able to take the plunge.

The goals outlined in The Strategy for Sustainable Farming and Food include to:

- Enable viable livelihoods to be made from sustainable land management, both through the market and through payments for public benefits

- Respect and operate within the biological limits of natural resources (especially soil, water and biodiversity)

- Achieve consistently high standards of environmental performance by reducing energy consumption, minimizing resource inputs and using renewable energy wherever possible

- Achieve consistently high standards of animal health and welfare

Timid though they are, these moves are very welcome. It is hoped that the support for organic and sustainable farming will increase in the future, at least to the levels that other EU farmers receive. This is very important if the goal of reducing imports of organic food to 30 per cent is to be met.

There is a final bit of good news for the UK. There are now new varieties of soya that can be grown in temperate climates. David McNaughton of Soya UK says that varieties bred for the UK at the University of Kiev could be guaranteed GM-free.[13]

Switzerland

The remarkable changes in agriculture in Switzerland (a member of EFTA) over the last few years should be a lesson for the rest of Europe. The country's unique form of democracy involves referenda on a variety of subjects, both at cantonal and national level. In 1996 there was a referendum on amending agricultural law in order to alter and extend existing subsidies aimed at ecological practices. Three tiers of support were proposed:

> Tier One provides support to traditional alpine grasslands and meadows, high-stemmed fruit trees under which cows and sheep can graze, and the protection of hedges.

Tier Two supports integrated farming practices which use reduced inputs, such as inorganic fertilizers etc., and higher ecological standards than conventional farming practice.

Tier Three provides the most support for organic farming.

Some 78 per cent voted in favour of the changes! To obtain the grants, farmers have to follow exacting rules and provide evidence that they have met several minimum conditions. These radical changes have already had a positive effect both on wildlife and on soil fertility. By the end of the 1990s, 85 per cent of farmland complied with the basic ecological standards. Some 5,000 farms are now fully organic and all farmers are set to fulfil the minimum standards. Pesticide applications have fallen by 33 per cent in the last decade, phosphate use is down 60 per cent, and nitrogen 50 per cent. Semi-natural habitats have expanded during the decade by 1–6 per cent in the plains and by 7–23 per cent in the mountains. Here is a country that has well and truly taken the plunge, and is well on the way to truly sustainable agriculture.[14]

There is much to learn from the Swiss experience. As we have seen from the consumer-led increases in organic sales in Western countries, the general public are often ahead of the politicians and orthodox experts. However, when central government and public opinion work together, mountains can truly be moved.

USA

In the single-most ambitious agricultural initiative of this century, Anthony Rodale, chairman of the Rodale Institute, declared a broad new vision for organic farming in the USA. At a trade show in Austin, Texas on 31 May 2003, where he collected the Organic Trade Association's (OTA) Leadership Award, he addressed leaders of the US organic industry. He called for 100,000 farmers, about 5 per cent of America's total, to make the transition to certified organic status by 2013. Currently, only 12,200 farmers in the US are certified organic.

The Massachusetts-based OTA reports that the market for organic food and products is more than $11 billion annually, and growing an estimated 20–30 per cent per year. Rodale, whose grandfather coined the term 'organic' in 1942, called on leaders to pool resources and work together towards a common, quantifiable vision. He continued:

"Are we willing to sit back and witness the degradation of our environment, and quietly lose control of the quality of food we eat every day? Organic farming is a good business decision for farmers. It's time for the US Department of Agriculture to affirm and promote organic as an economically viable form of agriculture." [15]

To achieve this vision, Rodale officially announced the launch of www.NewFarm.org, the nation's first organic farmer-orientated information centre, which provides articles, marketing tools and expert resources on crop and livestock production, direct marketing, and timely policy issues. It is the modern incarnation of *The New Farm Magazine*, the widely read print publication started in 1979 by Rodale's father. One of NewFarm.org's new, free services is its Organic Price Index, the first and only service offering farmers critical up-to-date pricing data. It also features:

- on-line training courses, teaching basic and advanced organic agricultural techniques

- success stories about innovative farmers and ranchers all across the USA

- scientific reports, and news related to the economics, politics, social issues and environmental impact of agriculture

- first-person accounts from around the world, including Cuba, Guatemala, India, Japan, Kenya, Laos, New Zealand, Thailand and Switzerland

In a separate initiative, the University of California is running a scheme teaching students how to grow organic food on plots provided by the faculty, and how to cook it in delicious ways. They have also replanted many of the ornamental gardens in the grounds on campus with edible fruit bushes and herbs which the students are encouraged to help themselves to.

On October 21st 2002 the US national organic standards were fully implemented, and in 2003 the US Department of Agriculture (USDA) finally promulgated a new 'USDA Organic Label'—an official seal of approval that assures shoppers that the food sold with this label is guaranteed to be grown without toxic chemicals, artificial fertilizers, antibiotics, growth hormones, GMOs or irradiation.

When this scheme was first mooted, the major food giants vehemently opposed it. They then changed tack and began instead to lobby the government to include GMOs, irradiation and even the use of toxic sludge as a manure under the definition 'organic'. The USDA has had more protests against this suggestion than any other federal agency has ever received on any previous proposed rule; it therefore had to back down, and the new labelling scheme was introduced.

In 2003 in Maine a legal fight between Oakhurst Dairy and the Monsanto company became nationwide news, with local newspapers claiming they were inundated with support for the local milk company, which sells its milk with the label "OUR PLEDGE—NO ARTIFICIAL HORMONES". In fact Oakhurst had support from many quarters around the world. Monsanto threatened to sue Oakhurst because, they argued, milk from cows injected with RBST hormone is no different from milk without this procedure. In August 2003 Ralph Nader, the leader of the US Green Party, stepped into the ring by offering free legal help to Oakhurst Dairy, because it was being financially outgunned by the international company Monsanto. Unfortunately the battle proved too much for Oakhurst, and a compromise was reached. Oakhurst Dairy was allowed to retain its slogan about no artificial hormones in their milk, but at the bottom of their labels is now a disclaimer stating that "FDA states: No significant difference in milk from cows treated with artificial growth hormones."

Stanley Bennett, who runs Oakhurst, an $85 million milk processor, says "We are in the business of marketing milk, not Monsanto's drugs. We in New England are into purity. The FDA may not have a problem with artificial growth hormones, but many consumers do." [16]

This is the latest chapter in a battle that has been raging for more than a decade. These hormones (RBST) are injected into the cows to increase milk yields, but there are many doubts about the health implications of the practice for consumers, such as premature puberty or even cancer.

Lawsuits have already forced the repeal of Vermont's 'Hormone-Disclosure Law' as well as stopping dairies in Illinois and Texas from marketing their milk as RBST-free, and the Federal Department of Agriculture, who support the use of RBST, have warned producers in Florida, New York, New Jersey and Minnesota against using labels that say "hormone-free" or "no hormones".

The production and sales of organic products in the United States is continuing to grow at a healthy pace, with sales of organic food and beverages increasing from $11 billion in 2002 to $13 billion in 2003, and the market research firm Datamonitor projects US organic sales to reach $30.7 billion by 2007.

However, it is very difficult to get a clear picture of the land and farms in organic production. Each state has its own figures, and there are around 53 registered certification bodies countrywide. To give a taste: in California the California Certified Organic Farmers (CCOF) said the acreage it certified increased by 20.5 per cent during 2001, 8.8 per cent in 2002 and 13 per cent in 2003, with 68,000 ha (170,000 acres) certified to date. In Vermont, certified acreage has grown from 23,638 in 2001 to 24,351 in 2002 and 30,387 in 2003 and the Washington Department of Agriculture, the largest certifier in the state, showed a growth of 7 per cent between 2001 and 2002 from 32,031 acres to 34,209 acres.

China

There are indications that China is starting to rediscover its organic and sustainable past. For years the country has taken to modern scientific farming in a big way, including GMOs. A whole generation of agricultural scientists and farmers have been schooled in monocropping with modern varieties and the use of fertilizers and pesticides. However, there are moves towards more sustainable ways of farming. As we have seen, Chinese scientists have been experimenting with growing more than one variety of traditional rice in one field. The mixed rice not only outcropped the modern single variety but disease was reduced by 94 per cent. As a result they were able to grow this mixture of rice strains without the use of pesticides, whilst at the same time enjoying an 18 per cent increase in production.[17]

In Fujian Province some 36,000 ha (89,000 acres) have azolla growing on the surface of the water in the paddy fields. As we have seen, azolla both smothers weeds and provides nitrogen for the rice when harvested and composted, or used as a green manure. It can fix more than 400 kg of nitrogen per hectare (350 lb per acre). There are also fish living in the paddy fields, which live off the azolla as well as providing manure, which also helps to feed the rice and provides more nitrogen than the azolla alone. The movement of the fish increases the oxygen content of the water, which helps to promote root growth. So the farmers have both fish as well as a slightly increased rice yield.[18]

The demand for organic products within China is growing, outstripping supply. Organic shrimp farming is growing largely for export, whilst organically produced poultry and pork are available in select locations in the domestic Chinese market. Certified Chinese organic products are increasingly seen in Malaysian and Singapore markets. It is predicted that China and other countries in the area stand on the threshold of an organic development wave.

India

The Indian government has been a little slow in realizing it has a huge potential to supply the growing organic food sector around the world, but things are starting to move. In 2002, at the International Conference on Organic Products in New Delhi, the Agricultural Minister, Ajit Singh, said that the government had formulated a national plan to ensure the production, promotion, market development and regulation of organic farming in the country, and would be providing the equivalent of US$19 million to start the project off. He said a national body would be set up under the project to formulate national standards of organic farming, and accreditation and certification agencies would be appointed to certify organic produce and train the farmers. Rajiv Pratap Rudy, a junior minister for commerce, said: "India has immense potential and core competence in organic foods which had become very popular in the Western world and can emerge as a major supplier to them." Fifty model organic farms would be established and assistance would be provided for setting up commercial production units for organic crops and inputs.

The Spices Board of India has initiated programmes to develop organic spice cultivation in the north-eastern parts of the country, because the international market is increasingly looking for sources of organically grown spices. This area is particularly suitable for growing pepper, large cardamom, turmeric and ginger; turmeric and ginger are already grown inorganically. The demand for organic pepper is the fastest growing. There is a shortage of quality warehousing and cold storage infrastructure, however, so the Spices Board is investing in these facilities, as well as training farmers in organic cultivation and setting up worm-composting units around the region.

Madhya Pradesh is the main area for growing soybeans, and in 2002 the state Co-operative Oilseeds Growers' Federation gave over 3,523 ha (8,700 acres) in the Malwa region for the contract growing of organic soya. It chose those plots that had already been used for growing organic cotton, and which had therefore not had any applications of chemical fertilizers and pesticides for a minimum of two years.

In July 2002, Adarsha Krishi Sahakari Kharedi Vikri Prakriya Sanstha launched the organic farming of cashew and coconut with an eye on the international market, the first co-operative body to do so. The Chairman, Prakash Velip, said that the society has so far trained more than 450 growers in organic farming; several queries had been received from abroad in the last three years for the supply of organic cashew nuts. It has also pioneered organic farming in Goa, and processed more than 100 tonnes of cashews at its factory at Koinamol in Sanguem Taluk in the previous season.

Organic sugar cane production is being promoted by a division of Transcorp Technologies and three leading sugar mills in Bangalore. The potential for organic sugar in the area is immense, owing to the growing international market and the fact that averages of conventional yields have fallen over the years. The waste from the mills is decomposed using cultures and enzymes for use as an organic manure to help grow the cane. The three mills produce 15,000 tonnes of this organic manure per year. Moreover, they have reported a higher extraction rate of sugar over the conventionally grown cane. Each mill takes cane from around 24,000–28,000 hectares (60,000–70,000 acres).

It is estimated that 65 per cent of the farmed land in India is already organic because traditional organic methods have never stopped. Here is a chance to encourage traditional farmers to stay organic and benefit from the growing markets around the world. The use of chemical fertilizers in eastern and north-eastern parts of the country is comparatively low, and yet there is plenty of food being produced. So far only 1,426 farms have been certified as organic, producing around 14,000 tonnes of produce per year, but this is a gross underestimate when all the traditional farms are taken into account. India is at last waking up to the huge potential for the sale of organic food both abroad and at home.[19]

A three-year project was launched in 2004 by the Research Institute of Organic Agriculture (FiBL Switzerland) to promote the Indian organic market at home and abroad using public relations, market research and pilot projects. The plan is to encourage networking between producers, processors, dealers and sellers. Also Naturland, the second largest organization in Germany, which promotes the certification of organic farms around the world, is helping accredited Indian organic farmers to export their produce. These include tea, coffee, spices and condiments. They will offer their members and partners a complete range of organic services including training, project set-up, certification and trade links. Smallholder Grower Group Certification is Naturland's specialty that enables small and marginal farmers to access certification. Both Naturland and the European Union have co-funded the construction of an organic tea factory in Kerala, where over 1,000 small tea growers are now processing organic tea in their own factory and marketing it.

And finally to finish on an exciting note: Mizoram is set to become the first state in India to ban the use of fertilizers and pesticides in agriculture and to encourage organic food production. The Governor of Mizoram, A. R. Kohli, told industrial leaders on 19 December 2003, "We will be the first state to go for organic food production. Once we do, other states in the region will follow." [20]

Africa

Certified organic production in Africa is very undeveloped, and where it is established it is usually aimed at exports, with the exceptions of Egypt and South Africa where there is an established home market. Tunisian production is largely aimed at exports, having its own organic (EU-compatible) standards, certification and inspection systems. Figures are very difficult to obtain, but there is evidence that around 50 per cent of African countries have certified organic farms, comprising around 320,000 hectares, half of which are in Uganda. There is also evidence of substantial recent growth of certified organic production in Ghana, Ethiopia, Tanzania and Zambia. Certified production schemes take two distinct forms: relatively large farms or plantations under single ownership, and smallholder groups, often supported by development aid.

Although certified organic production remains comparatively low in Africa, there is a much more important trend in parts of the continent—the development of a far larger agro-ecological movement, where small and subsistence farmers are turning to organic methods as a way of improving productivity and addressing the very real problems of food security for themselves and their families. Often these developments are encouraged by NGOs. These initiatives are designed to:

- Maintain and enhance soil fertility
- Combat desertification
- Promote tree-planting and agroforestry
- Develop low and no input ways of combating pests
- Promote the use of local seed varieties

- Maintain biodiversity
- Support the most vulnerable groups (particularly women and households headed by women), and
- Combat global warming,

Fine examples of projects in Africa where the principles above are applied, for both self-sufficiency and production for the sale of produce, are Harvest Help and GOAN. Harvest Help is a UK charity which helps farmers in countries like Zambia and Malawi to increase their food production using organic methods. As in many Third World countries, the costs of fertilizers and pesticides are too expensive. Instead they are now making their own compost and are intercropping their maize with plants like velvet bean and sunhemp, both of which help to increase soil humus and replenish soil nutrients. Many farmers have experienced dramatic increases in yields, whilst feeding their families, as well as preserving the environment, earning a living, and becoming more self-reliant.[21]

The Ghana Organic Agricultural Network (GOAN) was set up in 1995 with the support of the UK's leading organic horticultural organization, the Henry Doubleday Research Association. It is a demonstration and information centre for the dissemination of practical help and knowledge for those who wish to become organic farmers. Like many such organizations, it has the local expertise to apply organic theory to the local conditions, integrating traditional practices with new organic research—something that can only be carried out by local experts. It has set up a resource centre which provides an advisory service, runs training courses and farm demonstrations, and promotes organic agriculture in schools and to the general public. Through newspapers, radio and television, it has managed to raise awareness of organic agriculture throughout Ghana.

Farmer Field Schools (FFS) provide practical courses for farmers, with training in the use of organic techniques such as composting, the use of natural pesticides, mulching and manures. There has been a high uptake. Workshops are also held regularly for students and staff from local schools and colleges, which also attract representatives from government departments, village chiefs and other officials. GOAN has not only helped many farmers to go organic and improve their livelihoods, but is having a major influence on government agricultural policies.[22]

Cuba

The dramatic and inspirational changes in Cuba's food production over the last decade is a clear example of an organic 'mammal' developing unnoticed in the shadow of the conventional 'dinosaurs'. Up until the end of the 1980s Cuba was heavily dependent on the support of the Soviet Union, which was pledged to underpin the regime against US sanctions. In 1990, when the Soviet Union collapsed, the country was importing 57 per cent of all the calories it consumed each year. All of its wheat was imported, 90 per cent of its beans, 94 per cent of its fertilizers, 82 per cent of its pesticides and 97 per cent of its animal food. Its main export is traditionally sugar, for which the Soviet bloc paid three times the world price. In 1990, all this changed overnight. The loss of Soviet support was potentially catastrophic. The government could have chosen several different approaches, but what it decided to do was to declare an 'Alternative Model' for farming.

Faced with a cut in petroleum imports by half in two years, the response was to initiate a breeding programme for oxen to replace tractors. Monocropping has largely been replaced by diversification, crop rotation, green manuring, intercropping and soil conservation techniques. The integrated pest management (IPM) schemes to help replace pesticides, along with some 220 village-based small factories, produce various forms of biopesticides and biological controls, including 1,300 tonnes per year of *Bacillus thuringiensis* for the control of cabbage-white and other caterpillars, 780 tonnes a year of Beaveria sprays against beetles, 200 tonnes a year of Verticillium for white-fly control and 2,800 tonnes of Trichoderma, a natural enemy of pests. Sweet potato weevils are controlled by ants, which are attracted to the crop by cut banana stems baited with honey. Needless to say, many of these alternative control methods have proved to be more effective than pesticides. Composting has become the norm, with 173 worm composting units having been set up, producing 93,000 tonnes annually.

The main driving force for these changes has been the Grupo de Agricultura Organica (GAO), which has been instrumental in bringing together farmers, field managers, field experts, researchers and government officials to convince them that organic methods can truly provide sufficient food for Cuba's needs. Of course there has been resistance and scepticism from some of the scientific establishment, and from some of the orthodox farmers, especially the larger-scale ones, but they are making progress as improved yields and successful disease control provide proof of the new strategy.

Some press reports have suggested that Cuba has gone completely organic, but as yet this is not entirely true. To date, only the urban farming schemes can be said to be truly organic, but there is a growing trend towards sustainability. The eventual aim is to make all Cuban agriculture sustainable and already an estimated 200,000 hectares have been converted to semi-organic or sustainable production, with some 10,445 hectares truly organic. The other important part of the food production initiative is the intensive organic urban horticultural programmes that have been initiated. There are three types:

1. *autoconsumos*, self-provisioning gardens in schools and workplaces
2. *organoponicos*, raised container-bed gardens
3. *huertos intensivos*, intensive community gardens

These projects are inspirational in involving and galvanizing the urban population to become involved in their own food production, and play a significant part in the total food produced in the country. *Autoconsumos* are brilliant schemes for making all schools and workplaces as self-sufficient in food as possible, involving urban children and workers and exposing and involving them directly in food production, something of which most of them had no experience. *Organoponicos* can be used by individual householders and families where they have access to land, and *huertos intensivos* are shared and worked by urban dwellers that do not have access to their own plots. Together they produced some 4,200 tonnes of food in 1994, rising to 727,000 in 1999.

There were inevitably food shortages and widespread hunger in the first few years after 1990, as the new plan took time to get off the ground, and with the transformation of soil fertility that naturally takes place before a chemical-dependent system can become a thriving organic one, but by the end of the 1990s, available food was actually slightly above 1990 levels, confounding most of the sceptics.

The transformation of agriculture in Cuba was born out of necessity, but the country has turned a potential disaster into a success story that can provide a blueprint and inspiration for other countries. Increasingly governments, farmers and communities will have to make these kinds of changes.[23]

Mexico

It needs to be said that Latin America has a very ancient organic farming tradition, going back millennia, as did the East. The rotations of crops, variety selection, the creation and maintenance of soil fertility using composting and mulching, sophisticated irrigation systems, long-term planning and community land management were all features of American agriculture in the past. For example, the Aztecs in Mexico had a complex and intensive system of food production that involved irrigation from the mountains, elevated floating beds and regular crop rotations all over the lake surrounding their gigantic capital Tenochtitlan.

Today, Mexico leads the world in organic coffee production, with a total output of around 2,588 tonnes per year, followed by Guatemala, Peru, Kenya, Nicaragua, Tanzania, Brazil, Ethiopia, India and Madagascar. Organic certification for Mexican coffee began in 1962, and it is drunk in the US, Germany, Switzerland, Japan, Italy, Denmark, Spain, France and the UK.

According to the Mexican Coffee Council, the organic coffee growers are mostly indigenous farmers in Chiapas, Oaxaca, Veracruz and Guerrero states. Although the organic coffee trade represents just 1 per cent of global sales, for the Mexican growers it means healthy profits. Some 20,000 small farmers benefit from the higher price of their organic product.[24]

In Mexico, organic farms only constitute less than 1 per cent of the total, but the growth rate of all organic produce has been most spectacular recently, with a 442 per cent growth rate between 1996 and 2000 and then almost doubling between 2000 and 2002. In 2002, organic coffee constituted 69 per cent of all organic production, but they also grow organic maize, sesame, soybeans and other beans, a wide range of vegetables, nuts and fruit.

Honduras

More than 100,000 families make their living in the coffee industry, and because of the continuing surplus of coffee around the world they are beginning to suffer a 30 per cent decline in production. In contrast, in the Marcala Mountains, Lenca Indians are keeping hopes alive for a better life by growing organic coffee. On their small parcels of land they have planted special coffees that have captured the attention of the international market with their high quality and exotic flavours. Throughout the 145 ha (360 acre) plantation, they have incorporated other plants that provide much-needed shade for the coffee plants and also provide parallel crops through the growing season. Though it is still a small part of Honduras's annual production, it is hoped that it will grow in future as conventional coffee growing continues to suffer.[25]

Argentina

Organic farming has grown faster in Argentina than any other country in South America, where the area devoted to producing organic food has increased to just under 3 million certified hectares in 2000. The area devoted to producing organic food has increased from 5,000 to 300,000 ha (12,000 to 740,000 acres) in the last five years. Some 75 per cent of organic produce is sold to EU countries and 15 per cent to the US and Japan. In the province of Misiones, some 800 smallholders have been producing organic sugar, for which there is a growing demand from industries that produce organic yogurt, sweets and ice-cream. The producers of Misiones have signed a three-year contract to sell their entire production of sugar (12,500 tonnes) to the USA at a price 35 per cent higher than the normal world price. Argentina also has more land under zero-tillage, at 30 per cent, than any other South American country. Although not all of this land is run organically, this is still a major step forward.[26]

A large part of Argentina's organic production and exports is beef and Patagonian lamb. In 2002 there were 750,000 sheep and 122,000 head of cattle organically certified in the country.

Brazil

Brazil has for a long time been almost synonymous with the wholesale destruction of the rainforest, the unbridled pillage of natural resources, the loss of thousands of rare species of birds, animals and plants, and the genocidal treatment of certain indigenous peoples. So it is good to be able to report positive trends from this country.

Brazil is at the forefront of sustainable and organic farming. Fourteen million ha (35 million acres) of arable land is now under zero-tillage production. In the state of Rio Grande do Sul, 95 per cent of arable land has become zero-tillage in the past decade, and sustainable farming is heavily promoted. The farmers commonly use sunhemp, which has the remarkable property of having 3 metre (10 ft) deep roots that mine the subsoil and help to rejuvenate the land. On the bigger farms they use soya beans instead of mucuna to improve the soil. In Santa Catarina state, the Government has gone one step further by throwing down a challenge to make it the first to be free of the use of agrochemicals.[27]

In 2001 Brazil had 275,576 certified hectares of organic production; by 2003 this had risen to 800,000 hectares. The growth in organic agriculture, certified and non-certified, in Brazil is around 50 per cent per year and is contributing to the creation of a diversified production chain. Three Indian communities in Mato Grosso, Mato Grosso do Sol and Acre states are making profits with the production and commercialization of honey, mango and urucum. The urucum shrub has seed which produces the natural food colour anatto. The Yanawama community alone produces 3 tonnes of urucum per year with an income of US$81,000. There is also a growth in the production of herbal products from plantations of the andiroba tree (*carap guianensis*). Its properties are analgesic, antibacterial, anti-inflammatory and insecticidal, and it is becoming increasingly popular around the world.[28]

The most amazing story to come out of Brazil in the last few years, however, concerns *terra preta*, and could have profound and far-reaching benefits for organic and sustainable farming not only in South America but around the world, particularly in tropical countries where the land suffers from high rainfall and low natural fertility. Over 65 per cent of the world's rainforests, and around 75 per cent of the Amazonian forest, can be considered as 'wet-deserts', because they grow on red and yellow clay-like laterite soils which are rich in aluminium oxide and iron oxide, but little else. As a result they are very acidic, and naturally low in nutrients, and are unable to bind and hold plant nutrients. The reason is that they are both very old and impoverished, largely because of a lack of volcanic activity over a long geological period. As a result, no new nutrients have been brought up to the surface and the soils have become so weathered that they are largely devoid of minerals such as phosphorus, potassium, calcium and magnesium, all of which are essential for good plant growth. These huge and prolific jungles only exist because most of the nutrients and carbon are locked up in the vegetation, which in the hot and humid climate is rapidly recycled as soon as it dies. When the forest is cut down and the land used for agriculture it grows vigorous crops for only one to four years, after which there is a catastrophic decline in fertility, which can only be maintained by copious amounts of fertilizers.

However, there are areas of the Amazon which, against all expectations, have a rich, deep, dark and very fertile soil known as *terra preta* ('black earth'). Strangely, this soil consists of some 30 per cent black carbon as well as high levels of humus and massive populations of mycorrhizae and bacteria. To add to the mystery, these areas are also littered with prehistoric potsherds, indicating settled human habitation over a long period in the past. The area covered by the *terra preta* is huge—twice the size of the UK!

The Organic Uplands: A Vision of the Future

In 1542, the Spanish Conquistador Francisco de Orellana travelled up the Rio Negro, one of the Amazon's great tributaries, in search of El Dorado. Here he saw huge walled cities and a network of villages and farms. When he returned home, no one believed his story. When further expeditions arrived seventy-five years later, they found no civilization, although they did find an extensive network of raised causeways with canals running alongside. As a result de Orellana was branded a liar; it was believed that the Amazon soils were found to be so poor that they could not possibly support the kind of civilization he had described. What probably happened, however, is that the civilization he discovered suffered a catastrophic collapse in population between his discovery and future expeditions, as a result of the introduction of European diseases such as smallpox and measles by Orellana and his fellow explorers.

Recent research has uncovered this lost civilization, which has been estimated to have existed for about 2,000 years. The people had managed to overcome the problem of the poor soil. They discovered that when they cut down the trees to create their fields, if instead of using the usual cut-and-slash methods, they carefully burned the wood to make charcoal, and then crushed and incorporated it into the soil, they could improve fertility. The charcoal, in conjunction with high levels of organic matter, was the key to their success maintaining a permanent and sustainable organic agricultural system for so long.

The charcoal appears to play the same role as clay minerals in 'locking up' excess nutrients until they are needed by the plants, except it is far more effective in doing so: it lasts for hundreds, possibly thousands of years, continuing to do its job all the time. Scientists have added both charcoal and fertilizer to poor rainforest soil, and found that yields on the charcoal plots were 880 per cent higher than those which only had fertilizer.

There also seems to be another remarkable feature of the *terra preta*: it can reproduce itself. When left undisturbed it grows, probably because of the huge population of bacteria and fungi processing the leaves and dead vegetation that fall onto it over the years in a highly efficient way. Scientists are now busy trying to understand and re-create this biological 'cocktail', which could help to breathe life back into depleted and poor soils around the world, helping to feed people on marginal land and even to revitalize soils that have being denuded by modern farming methods. Once again we have lessons to learn from the past.[29]

Zero-Tillage and the Mucuna Bean

Some of the most exciting developments in sustainable farming techniques and practices are happening in Central and South America. Although the farms are often not organic, especially the larger ones, nonetheless the more sustainable practice of zero-tillage has grown at an almost exponential rate in the last few years. In Guatemala and Honduras zero-tillage and organic practices have been particularly successful, especially on the steep, easily eroded hillsides with depleted soils because of the introduction of the mucuna (or magic) bean into the farmers' rotations: the combination of growing the mucuna bean along with the adoption of zero-tillage techniques has been the key to success. First the beans are sown on the depleted land, resulting in prodigious growth, quantities of nitrogen fixed by the roots and yields of 100 tonnes per ha (40 tons per acre). When the plants have died at the end of the season, maize is planted straight into the bean residue, without ploughing or cultivation. After this first green-manure crop of mucuna beans, crops of maize and beans are sown together. The soil comes back to life, gaining humus, with high bacteria and mycorrhizae activity, and an increase in worm population. The bean plants can also be harvested to make high-quality compost. Yields of grain have doubled or trebled using these methods. The beans are also eaten; they are often made into flour and mixed with wheat and eggs for a high-protein bread. Molacus and chicuna beans are also used in the same way to produce rich and fertile soils.

These farmers are helped by organizations that provide knowledge and practical help to become sustainable, and the governments are also encouraging this trend. The farmers get better prices for their organic produce, and this, combined with the improved yields and greatly reduced inputs, is at last providing a reasonable living for many who had given up hope.[30]

The Organic Revolution

Science and Spirituality

There are those who still argue that, to become totally acceptable, above all to the agricultural scientific elite, organic and sustainable farming must be understood entirely in Western scientific terms, whilst at the same time marginalizing or ignoring those in the past, mostly from non-Western cultures, which arrived at organic husbandry through a combination of practical, empirical, philosophical and spiritual routes. Most Western organic pioneers, although dedicated to scientific understanding, were also religious or spiritual personalities. If one does not take this into account, it is not possible to understand the true motivations of those from both Western and non-Western cultures who have farmed organically. To try to understand organic farming exclusively through modern Western scientific eyes is to throw out the baby with the bathwater. I would venture to suggest that one of the reasons that organic farming and food are becoming increasingly attractive today, is precisely because there is more than science underpinning them, however important that is.

None of this in any way implies a rejection of science, which has transformed our understanding of soil ecology and biology, and which has taken organic and sustainable approaches to growing food to new heights. Nor should one dismiss the many genuine scientists who have sought to understand and reveal Nature's mysteries. Awe and love of Nature is not confined to green enthusiasts and ecologists, but provides the driving force behind many scientists' work. The organic approach to farming includes scientific understanding in the broadest sense, taking the best from more than one tradition, culture and period of history and recognizing that all peoples of all ages provide valuable insights into ecology and farming. As Vandana Shiva commented in her book *Biopiracy*,

> "Science is an expression of human creativity, both individual and collective. Since creativity has diverse expressions, I see science as a pluralistic enterprise that refers to different 'ways of knowing'. For me, it is not restricted to modern Western science, but includes the knowledge systems of diverse cultures in different periods of history. . . . Recognition of diverse traditions of creativity is an essential component of keeping diverse knowledge systems alive. This is particularly important in this period of rampant ecological destruction, in which the smallest source of ecological knowledge and insights can become a vital link to the future of humanity on this planet.[31]

The more organic husbandry is scientifically studied, the more it is shown to work. The more research the better, and the more we understand the biology, the ecology and the chemistry of the processes of plant nutrition, the more organic and sustainable agriculture will become. Indeed one would be a fool not to embrace scientific understanding of the subject. But to see organic husbandry exclusively in terms of the existing, limited scientific paradigm is to fall into the same trap that has been largely responsible for the problems we have with modern farming in the first place.

Organic Growth

One of the exciting and impressive aspects of the organic revolution is how, in many cases, it is coming from the bottom up. In Europe, the USA and many other areas, the growth in organic farming has largely been driven by consumer demand—which continues to lead supply—while in other countries, as in India, Africa and South America, it has been driven by the farmers themselves, especially small farmers, because they get better results and a better living. In other areas of the world the changes are spearheaded by governments or local institutions, as in Cuba, Indonesia and China. What is important, however, is that it is happening, as more and more people realize that in the end it is the only sensible course.

As long as organic standards are maintained, and as long as there is consumer confidence in the labelling and inspection systems, especially with products from other countries, the growth in organic produce will continue. In a rush to fulfil the growing market, however, there is a fear that standards could be lowered, so there is an increasing need to keep vigilant. This is something that concerns both organic producers and the buying public. As Nicolas Lampkin has said, "It will be essential . . . that the underlying holistic, ecological perspective is not lost sight of due to market and other pressures."[32]

Transition

The world of agriculture is in transition, with the old paradigms and practices apparently well entrenched, but with a small but growing sustainable and organic sector alongside them. At such times, the conventional view and practices seem unassailable, as the dinosaurs appeared to be in their day. But when we look closer, we can see that cracks are appearing, and conventional farming is in crisis. Fritjof Capra described the mechanics of this transitional process very succinctly:

> "During the process of decline and disintegration the dominant social institutions are still imposing their outdated views but are gradually disintegrating, while new creative minorities face the new challenges with ingenuity and rising confidence. . . . While the transformation is taking place, the declining culture refuses to change, clinging ever more rigidly to its outdated ideas; nor will the dominant social institutions hand over their leading roles to the new cultural forces. But they will inevitably go on the decline and disintegrate while the rising culture will continue to rise, and eventually will assume its leading role.[33]

There are many different reactions by the establishment to the new creative ideas. First there is dismissiveness, then ridicule, then a partial attempt at assimilation whilst attempting to preserve the bulk of the old ideas—what the President of the Soil Association, Jonathan Dimbleby, described as 'greening' the process at the edges. Finally when they really become rattled they go onto the attack. We are seeing a combination of all these at the moment.

Sir John Krebs, the former head of the Food Standards Authority, dismisses the growth and interest in organic food by suggesting it is a passing fad. The UK government wants to encourage organic farming and some green projects, but within the context of the 'real' farming sector, which their advisers see as the mainstream. Then there are those who have decided to take the bull by the horns and launch an all-out attack. It appears from their own utterances that they see anything 'green' as a threat to the capitalist system and even to modern civilization as we know it. Denis T. Avery falls into this category. He argues that it is essential to use biotechnology, pesticides, irradiation, factory farming and free trade, and it is the 'greenies' and 'organic frenzies' who are threatening the world with famine and loss of habitat for the wildlife they love so much, because if we farm without synthetic pesticides, petrochemical fertilizers and biotechnology, we will not be able to feed the world.

It is these promoters of 'modern, scientific' farming and GM technology who feel most threatened, although it is not always easy to understand why. Their sometimes fanatical rhetoric sends chills up the spine. For example, the pro-GM website www.AgBioWorld.org equates Vandana Shiva, Greenpeace and Friends of the Earth with those who bombed the twin towers in New York! It says:

> "With the recent attacks perpetrated by those who put political ideology above human life, we have an opportunity to re-evaluate the politics of Greenpeace and similar groups.... With the plane-bombs, we can see all of these groups for what they are. Perhaps after the crudescence of terrorism in New York and Washington, the world will wake up to the politically motivated destruction advocated by Greenpeace, Friends of the Earth and others, and see that this is wrong. Vandana Shiva has blood on her hands, so does Mae-Wan Ho. I recommend these folks lay low, very low, until political terrorism becomes fashionable." [34]

Organic farming is not going to go away. It is a truly creative, scientific and practical approach to producing food that will have to be willingly adopted wholeheartedly very soon, otherwise some very painful changes to our practices will be forced on us. Organic and sustainable farming has served many cultures very well over many centuries, whilst maintaining productive living soils. Many small farmers, usually women, have been carrying on highly intensive and productive sustainable practices over thousands of years, handing their knowledge down from generation to generation, and we have much to learn from them. However, organic husbandry is continuously evolving, as I have tried to show in this book, and as Nicolas Lampkin has commented:

> "Organic farming is not a fixed and static approach to agriculture; it is a dynamic, rapidly developing concept which is certain to evolve and adapt further over the next few years." [35]

Throughout this book I have stressed the point that we have a choice. We can perceive Nature as 'other', the enemy, something to be overcome, exploited or improved. Or we can recognize that it is not 'out there', but 'in here'. It is only when we can no longer tell where we end and Nature begins that we will truly change the way we grow our food and many other things besides. Only then will the destructive, life-damaging forms of agriculture that are in the ascendancy at the moment be overtaken by life-affirming practices. Then we will recognize that the choices we thought we had were illusory, and that in reality there was really no choice at all.

Notes

Chapter 1: What is Organic Farming?

1. Lampkin, *Organic Farming*, p.1.
2. King, *Farmers for Forty Centuries*.
3. Balfour, *The Living Soil*, p.24.
4. Howard, *An Agricultural Testament*, p.1.
5. King, p.252.
6. Shiva, BBC Reith Lecture 2000.
7. Brown, *The Shopper's Guide to Organic Food*, p.1.
8. Shiva, *Biopiracy*, p.106.
9. Lovelock, *Gaia*, p.ix.
10. *New Scientist*, Dec. 1996, p.28.
11. Lovelock, p.10.
12. Ibid. p.10.
13. Ibid. p.12.
14. Howard, p.1.
15. Lampkin, p.130.
16. Ibid. p.19.
17. *Encyclopaedia Britannica* 1963, vol 9, p.191.
18. Pretty, *Regenerating Agriculture*, p.3.
19. Ibid. p.6.
20. Bragg, 'In Our Time', BBC Radio 4, 10 July 2003.
21. Pretty, *Agri-Culture*, p.86.
22. Pretty, *Regenerating Agriculture*, p.70.
23. Howard, p.24.
24. Soil Association, *Living Earth* magazine, no. 210, Apr–Jun 2001, p.16.
25. Lampkin, pp.496-7.
26. Soil Association, 'Counting the Costs of Industrial Agriculture', 1996.
27. Brown, p.42,
28. Pretty, *Agri-Culture*, p.62.
29. Lampkin, pp.572-3.
30. Ingham in *Living Earth* magazine, Soil Association, Winter 2002, p.12.
31. Lampkin, p.584.

Chapter 2: From the Past to the Present

1. Lampkin, *Organic Farming*, p.3.
2. Conford, *The Origins of the Organic Movement*, p.15.
3. Pretty, *Regenerating Agriculture*, p.45.
4. Most of this information comes from Robert Temple's book *The Genius of China*, but some I absorbed from elsewhere.
5. King, *Farmers for Forty Centuries*, p.6.
6. Ibid. pp.190-2.
7. Ibid. p.45-7.
8. Ibid. p.79 and pp.185-8.
9. Ibid. pp.194-6.
10. Ibid. p.194.
11. Ibid. p.199.
12. Ibid. p.9.
13. Ibid. pp.209-10.
14. Ibid. p.251.
15. Ibid. Fig 69. p.131.
16. Ibid. pp.266-7.
17. Ibid. p.286.
18. Ibid. pp.208-9.
19. Ibid. pp.266-7.
20. Ibid. pp.34-5.
21. Ibid. pp.263-4.
22. Ibid. pp.266-7.
23. Ibid. p.347.
24. Ibid. p.360.
25. Ibid. p.73.
26. Ibid. pp.164-5.
27. Ibid. p.13.
28. Ibid. pp.275-6.
29. Lampkin. p.66.
30. *Encyclopaedia Britannica*, 1963, under 'Rotation of Crops'.
31. Most of the information about the Norfolk four-course rotation and much of the following section on the enclosures, comes largely from memory, due to a brilliant and enthusiastic history teacher at school called David Hughes, who taught us the period of British history 1760–1914, which included both the Industrial and Agricultural revolutions. (Details and dates from *Encyclopaedia Britannica*.)
32. *Encyclopaedia Britannica*, 1963, under 'Enclosure'.
33. Ibid.
34. Lampkin, p.128.
35. Ibid. pp.128-9.
36. Conford, pp.50-3.
37. Tompkins, *The Secret Life of Plants*, pp.205-9 and Conford, pp.53-7.
38. Ibid. p.205.
39. Howard. p.165.
40. Tompkins, p.208.
41. Soper, *Bio-dynamic Gardening* and *Star & Furrow*, for details of experiments etc..
42. Tompkins, pp.210-11.
43. Ibid. p.210.
44. Lampkin. p.31.
45. Conford, pp.100-3 and Tompkins, p.212.
46. Tompkins, p.212.
47. Conford, pp.103-8.
48. Conford, pp.108-10.
49. These are my own notes, having read Turner's books many times and used many of his ideas in practice.
50. Turner, pp.20-1.
51. See Elm Farm Research Centre Bulletin, which provides an up-to-date guide on the latest research results and policy developments.
52. BBC News Online: Sci/Tech (30 May 2002) and a summary of research in *Living Earth*, Soil Association, Winter 2002, p.4.

Chapter 3: The Living Soil

1. Howard, p.1.
2. Ibid. p.26.
3. Ibid. pp.27-8.
4. Ibid. p.28.
5. Ibid. p.26.
6. *Living Earth* magazine, Soil Association, no. 210, p.16.
7. Russell, *Soil Conditions & Plant Growth*, 11th ed., p.658.
8. *Living Earth* magazine, Soil Association, no. 210. p.16
9. Ibid. p.16.
10. Russell, pp.538-41.
11. Ibid. p.457.
12. Lampkin, p.78.
13. Ingham in *Living Earth* magazine, Soil Association, p.16.
14. Howard, p.25.
15. Lampkin, p.78.
16. Ingham in *Living Earth* magazine, Soil Association, p.16.
17. Ibid. p.16
18. Lampkin, p.65.
19. Darwin, Charles, *The Formation of Vegetable Mould, Through the Action of Worms*.
20. Lampkin, p.24 and Russell, p.509.
21. Lampkin, pp.24-6.
22. Russell, p.512.
23. Lampkin, p.24.
24. Ibid. p.65-6.
25. Russell, p.663.
26. Ibid. p.747 and Lampkin, pp.71-2.
27. Lampkin, p.72.
28. Ibid. p.76.
29. Ibid. p.76-8.
30. Howard, p.1.
31. Turner, *Fertility Farming*, pp. 53, 87.
32. Pretty, *Regenerating Agriculture*, p.111.
33. Pretty, *Agri-Culture*, pp.92-3.
34. Brown, p.44.
35. Ingham, www.soilfoodweb.com.
36. Lampkin, pp.242-3.
37. Ibid. p.248.
38. Ibid. p.248-51.
39. Pretty, *Regenerating Agriculture*, p.107.
40. Lampkin, pp.251-2.
41. Ibid. pp.313-17.
42. Turner, p.82.
43. Ibid. pp.80-1.
44. Lampkin, p.217.
45. Howard, pp.3-4.
46. Lampkin, p.129.
47. Ibid. p.129.
48. Ibid. pp.236-7.
49. Ibid. pp.232-3.
50. www.global-good-news.com (agriculture).
51. Lampkin, pp.219-20.

Chapter 4: Nutrition and Food Safety

1. Pretty, *Agri-Culture*, p.92
2. Interview with Sir John Krebs on 'Countryfile', BBC1, 12.30 3rd Sept. 2000 and news.bbc.co.uk Friday 1st Sept 2000.
3. *Living Earth*, Soil Assn., Spring 2003, p.25.
4. Brown, p.38 and Pretty, *Agri-Culture*, p.64.
5. Lampkin, pp.568-9.
6. *The Organic Way*, HDRA's magazine, Summer 2003, p.6, and the American Chemical Society website, www.acs.org.
7. Brown, p.127.
8. *The Organic Way*, HDRA's magazine, issue 176, Summer 2004, p.5 and http://icwales.icnetwork.co.uk.
9. 'Organic Farming, Food Quality & Human Health', Soil Association, pp.20-21, and Public Health Laboratory Service 2001, 'The Microbial Examination of Ready-to-Eat Organic Vegetables from Retail Establishments', PHLS Environmental Surveillance Unit, London.
10. Lampkin, p.563.
11. 'Organic Farming, Food Quality & Human Health', Soil Association, pp.72-86, which questions the validity of 71 out of 99 published papers on the subject, solely on the basis of scientific vigour.
12. Brown, p.39.
13. Pretty, *Regenerating Agriculture*, p.60.
14. Dudley, *This Poisoned Earth*.
15. Lampkin, p.558.
16. 'Organic Farming, Food Quality & Human Health', Soil Association, pp.18-19.
17. Brown, p.39.
18. Abou-Donia M.B. et al, *Journal of Toxicol. Environ. Health* 1996, no. 48, pp.35-65.
19. *The Organic Way*, HDRA, issue 168, Spring 2001, p.6.
20. Food Standards Agency (2002) Committee on Toxicity of Chemicals in Food, Consumer Products & the Environment: Risk Assessment of Mixtures of Pesticides & Similar Substances, FSA.
21. Brown, p.39.
22. Ibid. p.343.
23. Ibid. pp.47-48.
24. DEFRA Report 2001.
25. Pesticide Action Network UK, 'Pesticides in Water' Briefing 2000, p.4.
26. Brown, p.42.
27. Lampkin, p.578.
28. Ibid. pp.25-6.
29. Soil Association. 'Counting the Costs of Industrial Agriculture', 1996.
30. Pretty, *Agri-Culture*, pp.57-9.
31. 'Organic Farming, Food Quality & Human Health', Soil Association, p.23. Quote from WHO/73, 20 Oct. 1997.
32. Pretty, *Agri-Culture*, pp.64-65.
33. Lampkin, p.565.
34. 'Organic Farming, Food Quality & Human Health', Soil Association, p.23.
35. Brown, pp.10-11.
36. Pretty, *Agri-Culture*, p.93.

Notes

Chapter 5: Genetically Modified Foods

1. HRH Prince Charles, BBC Reith Lecture 2000.
2. John Fagan, *Genetic Engineering: The Hazards, Vedic Engineering: The Solutions*, pp.9-11.
3. Shiva, p.28.
4. Fagan, p.40.
5. www.newscientist.com.
6. www.wfp.org.
7. Pretty, *Agri-Culture*, pp.82-85.
8. Devinder Sharma, *New Scientist*, 8th July 2000, pp.44-45.
9. Vandana Shiva, BBC Reith Lecture no.5, 2000.
10. Howard, pp.192-3.
11. Fagan, pp.8-9.
12. Jean McNeil, 'Growing Concerns' in *Earth Matters* magazine, Summer 1999, pp.9-11.
13. Research Foundation for Science, Technology & Ecology (RFSTE), posted 3 Oct 2002. A-60, Hauz Khas, New Delhi 110016, India. Email: rfste@vsnl.com.
14. *Nature*, vol. 413, 27 Sept 2001.
15. To read the report see www.biotech-info.net/technicalpaper6.html.
16. Most of the information in this section can be found at: www.soilassociation.org GMOs, Tests for the Effects on Health.
17. *GM Nation? The Public Debate*, HMSO, and www.gmnation.org.uk.
18. BBC Radio 4, 'Farming Today', 25 April 2002.
19. *Living Earth* magazine, Soil Association, Winter 2002, pp.12-13.
20. Ibid. p.13.
21. Steven M. Druker, sdruker@biointegrity.org
22. *Living Earth* magazine, Soil Association, Winter 2002, p.13.
23. Ibid. p.13.

Chapter 6: Maintaining Soil Fertility

1. Howard, *An Agricultural Testament*, p.2.
2. Ibid. pp.39-52.
3. Ibid. pp.27-28.
4. Ibid. p.42 and Lampkin, p.100.
5. Ibid. p.49.
6. Lampkin, p.96.
7. Ibid. p.93.
8. Hills, Lawrence D., *Grow Your Own Fruit and Vegetables*, p.27.
9. King, pp.250-251.
10. Turner, *Fertility Farming*, pp.107-109.
11. Lampkin, pp.104-106.
12. Ibid. p.107.
13. Richard & Walker, *Composting Trends & Technologies*, Cornell University Education Department, USA, p.1.
14. The Composting Association, 2003-4, pp.6-7.
15. The Composting Association, 2003-4, p.36.
16. Richard & Walker, pp.1-2.
17. Ibid. p.3.
18. The Composting Association, p.14
19. United States Environmental Protection Agency (EPA), 'Innovative Uses of Compost', October 1997, pp.4-5.
20. The Composting Association, pp.44-45.
21. Ibid. p.27.
22. Ibid. p.27.
23. Richard & Walker, p.2.
24. The Composting Association, 1999, pp.6-7.
25. Ibid. pp.8-9.
26. The Composting Association, 2003-4, p.6.
27. The Composting Assn., 1999, pp.40-41.
28. United States Environmental Protection Agency (EPA). 'Innovative Uses of Compost'.
29. Lampkin, pp.92-93.
30. Pretty, *Agri-Culture*, pp.86-88.
31. King, pp.193-6.
32. Lampkin, p.119.
33. Richard & Walker, p.2.
34. Lampkin, pp.119-120
35. Pretty, *Agri-Culture*, pp.99-100
36. This information was obtained from a display on sewage treatment at the now defunct Earth Centre, Doncaster, Yorkshire.
37. Further information on these projects can be found by contacting: FuelCell Energy, Inc. Steve Eschbach, 203-825-6000, seschbach@fce.com, and Bourns College of Engineering – Centre for Environmental Research & Technology, the University of California, Riverside.

Chapter 7: Control versus Co-operation

1. Malik, *Man, Beast & Zombie*, p.123.
2. Marx, *Economic and Philosophic Manuscripts*, 1844, p.61.
3. Pretty, *Agri-Culture*, p.8.
4. Hartmann, Thom, *The Last Hours of Ancient Sunlight*, p.154.
5. Ibid. pp.154-5.
6. Ibid. pp.270-1.
7. Melvin Bragg, 'European Attitudes to Nature: From the Greeks to the Present Day', BBC Radio 4, 10 July 2003.
8. Malik, p.31.
9. Capra, *The Tao of Physics*, p.141.
10. There are many books on Chinese Art, for example O'Neill, Amanda, *The Art of Chinese Watercolours*, Siena.
11. Capra, *The Tao of Physics*, pp.149-50.
12. Melvin Bragg, BBC Radio 4, 10 July 2003.
13. Ibid.
14. Ibid.
15. Hartmann, pp.86-88.
16. Malik, pp.3-4.

17. Melvin Bragg, BBC Radio 4, 10 July 2003.
18. Malik, p.4.
19. Ibid. p.3.
20. Vandana Shiva, *Biopiracy*, pp.50-51.
21. Malik, p.34.
22. Shiva, pp.50-51.
23. Carolyn Merchant, *The Death of Nature*, 1980, p.169.
24. Melvin Bragg, BBC Radio 4, 10 July 2003.
25. Ibid.
26. Malik, pp.61-64.
27. Capra, *The Tao of Physics*, p.27.
28. Devinder Sharma, writing in *New Scientist* magazine, 8 July 2002, pp.44-45.
29. *Encyclopaedia Britannica* article on Descartes: 'The Philosophy of Descartes'.
30. Pretty, *Regenerating Agriculture*, p.13.
31. Kuhn, *The Structure of Scientific Revolutions*, pp.24-5.
31. Lampkin, pp.56-7.
32. Shiva, p.51.
34. Ibid. p.51.
35. Malik, p.54.
36. Pretty, *Agri-Culture*, p.29-32.
37. Melvin Bragg, BBC Radio 4, 10 July 2003.
38. Ibid.
39. Darwin, *Descent of Man*, vol.1, p.105.
40. Malik, pp.109-110.
41. Capra, *The Turning Point*, p.17.

Chapter 8:
The Challenges Ahead

1. Pretty, *Agri-Culture*, p.188.
2. Soil Association, Response to the 'Policy Commission on the Future of Farming & Food', October 2001, p.3.
3. Mayer Hillman, *How We Can Save the Planet*, p.7.
4. Ibid. p.18.
5. *Oil & Gas Journal*, and Dr Colin Campbell, *The Twenty First Century: The World's Endowment of Conventional Oil & its Depletion*, 1996.
6. Hartmann, p.21.
7. Campbell in Douthwaite (ed.), *Before the Wells Run Dry*, pp.30-40.
8. Lampkin, p.584.
9. Ibid. p.584.
10. Hartmann, pp.17-19.
11. Ibid. p.29.
12. Soil Association, Response to the 'Policy Commission on the Future of Farming & Food', October 2001, p.7.
13. Pretty, *Agri-Culture*, pp.75-6.
14. Paul Hawken, Amory Lovins & Hunter Lovins, *Natural Capitalism: The Next Industrial Revolution*, pp.32-7.
15. UK's Loughborough University's Centre for Renewable Energy Systems Technology (CREST) already have their Hydrogen & Renewables Integration system up and running at West Beacon Farm.
16. Scheer, Hermann, *A Solar Manifesto*, p.190.
17. DEFRA, 'The Strategy for Sustainable Farming & Food', Section 2.1.1., pp.20-1.
18. What has happened to these plans and whether they have been put into practice, I have been unable to find out.
19. Keith Hall (ed.), *The Green Building Bible*, pp.182-4.
20. Soil Association, Response to the 'Policy Commission on the Future of Farming & Food', October 2001, p.8
21. Ibid. p.7
22. HM Government. Report of the Policy Commission on the Future of Farming & Food, Section B, Chart 1: The UK food chain.
23. Capra, *The Turning Point*, pp.238 & 242.
24. Ibid. p.243.
25. Shiva, BBC Reith Lecture 2000.
26. Shiva, *Biopiracy*, p.81.
27. HM Government, Report of the Policy Commission on the Future of Farming and Food, p.A2.
28. Hawken et al, p.61.
29. Capra, *The Turning Point*, p.433.
30. Pretty, *Agri-Culture*, p.106.
31. Soil Association, Response to the 'Policy Commission on the Future of Farming & Food', October 2001, p.7
32. DEFRA. The Strategy for Sustainable Farming & Food. Section 2.4.1, p.33.
33. Lampkin, p.492.
34. Ibid. p.497, figure 13.2.
35. Ibid. p.498, table 13.6.
36. Hodges, R. D., *Who Needs Inorganic Fertilizers Anyway?*, 1978, p.6.
37. Lampkin, p.496, table 13.4.
38. DEFRA, 'Notice to Farmers and Growers in England', from the Office of Lord Whitty, Minister for Food, Farming and Sustainable Energy.
39. Pretty, *Agri-Culture*, pp.75-6.
40. For a good introduction to Permaculture I recommend Bill Mollison's book *Permaculture: A Practical Guide for a Sustainable Future*, 1990.
41. See 'The Centre for Environment & Society', The University of Essex and also Pretty, *The Living Land*, 1998.
42. Howard, p.1.
43. Ibid. p.35.
44. *The Organic Way*, HDRA, issue 171, Spring 2002, pp.31-3.
45. Refer to the 'Forestry Stewardship Council', which is the only truly independent certification scheme in the world, sup-

46. DEFRA, 'The Strategy for Sustainable Farming & Food', Section 2.5.4.
47. Pretty, *Regenerating Agriculture*, p.31.
48. Shiva, BBC Reith Lecture 2000.
49. Ibid.
50. Maurice Malanes, 'Traditional Rice Varieties Still Best for Terraces', in the *Philippine Daily Enquirer*, 28 May 2003.
51. Pretty, *Regenerating Agriculture*, pp.117-8.
52. Ibid. chapters 5 & 6.
53. Ibid. pp.204-6.
54. *The Organic Way*, HDRA, issue 172, Summer 2003, pp.42-3.
55. Brown, p.260.
56. *The Organic Way*, HDRA, issue 172, Summer 2003, p.42.
57. Ibid. issue 169, Autumn 2002, pp.18-19.

Chapter 9: The Organic Uplands: A Vision of the Future

1. This information originally came from an article in *The New York Times*, by Michael Pollan, an edited version of which was printed in the Soil Association's *Living Earth* magazine, Winter 2002, pp.22-24.
2. Lampkin, pp.353-357.
3. Pretty, *Regenerating Agriculture*, pp.29-31.
4. Ibid. p.7.
5. Shiva, BBC Reith Lecture 2000.
6. Devinder Sharma, 'The Green Revolution Turns Sour' in *New Scientist* magazine, 8 July 2000, pp.44-45.
7. Pretty, *Regenerating Agriculture*, p.75.
8. *New Scientist* magazine, 8 July 2000, p.45.
9. *The Organic Way*, HDRA, issue 168, Summer 2002, pp.18-19.
10. Most of these figures for organic trends around the world come from a comprehensive IFOAM report edited by Minou Yusssefi & Helga Willer through the Institute of Rural Sciences: go to www.organic.aber.ac.uk/statistics/index.shtml and follow their links. Also see the Soil Association's 'Organic Food & Farming Report' 2000.
11. Soil Association, Response to the 'Policy Commission on the Future of Farming & Food, October 2001, p.8.
12. Ibid. p.3.
13. *The Organic Way*, HDRA, issue 169, Autumn 2002, p.7.
14. Pretty, *Agri-Culture*, pp.75-6.
15. www.organics.com.
16. Margot Roosevelt Leeds. 'Got Hormones? The Simmering Issue of Milk Labels Boils Over when an Agrochemical Giant Sues Small Farmers in Maine', in *Time Magazine* (USA), 22 December 2003.
17. *Nature* magazine, August 2000.
18. Pretty, *Regenerating Agriculture*, p.22.
19. Most of the information in this section on India comes from www.organicts.com, a very useful site to use if you wish to keep up to speed with world organic events.
20. 'Mizoram to be Organic State' in *Times of India*, 18 December 2003.
21. *The Organic Way*, HDRA, issue 172, Summer 2003, p.6.
22. Ibid. pp.18-19.
23. Pretty, *Agri-Culture*, pp.74-75.
24. www.organicts.com.
25. Ibid.
26. Marcela Valente, 'Agriculture-Argentina: Organic Farmers Move in From the Margins', Inter Press Service, 15 September 2002.
27. Pretty & Hine, 'SAFE-World Research: 47 Portraits of Sustainable Agricultural Projects & Initiatives.' Centre for Environment & Society, University of Essex, 2001.
28. www.organicts.com.
29. 'The Secret of El Dorado', Horizon programme, BBC2, December 2002, and B. Glaser et al, *Organic Geochemistry*, 1998, 29, pp.811-819. Institute of Soil Science & Soil Geography, University of Bayreuth, Germany.
30. Julian Pettifer, 'The Magic Bean', BBC News, Radio 4, 8 June 2001, and Pretty, *Agri-Culture*, pp.49-50.
31. Shiva, *Biopiracy*, pp.13-14.
32. Lampkin, p.608.
33. Capra, *The Turning Point*, p.466.
34. *Living Earth* magazine, Soil Association, no.213. March 2002, p.7.
35. Lampkin, p.608.

Bibliography

BOOKS

Balfour, Eve, *The Living Soil & The Haughley Experiment.* New York: University Books, revised edition 1976.

Bromfield, Louis, *Malabar Farm.* Cassell 1949.
Out of the Earth. Cassell, 1951.

Brown, Lynda, *The Shopper's Guide to Organic Food.* London: Fourth Estate, 1998.

Capra, Fritjof, *The Tao of Physics.* London, Flamingo, 1975.
The Turning Point. London: Flamingo, 1984.

Carson, Rachel, *Silent Spring.* London: Penguin Books, revised edition 1999.

Conford, Philip, *The Origins of the Organic Movement.* Edinburgh: Floris Books, 2001.

Darwin, Charles *The Formation of Vegetable Mould, Through the Action of Worms.* London: Faber & Faber, 1966.

Douthwaite, Richard (ed.) *Before the Wells Run Dry,* Dublin, FEASTA, 2003.

Dudley, Nigel, *This Poisoned Earth.* London: Piatkus Books, 1987.

Fagan, John, *Genetic Engineering: The Hazards, Vedic Engineering: The Solutions.* Fairfield USA: Maharishi International University Press, 1995.

Faulkner, Edward, *Ploughman's Folly.* London: Michael Joseph, 1945.

Flowerdew, Bob, *Bob Flowerdew's Organic Bible.* London: Kyle Cathie Ltd, 1998.

Fukuoka, Masanobu, *One Straw Revolution.* New York: Bantam Books, reprint 2002.

Hall, Keith (ed.) *The Green Building Bible.* Llandysul, Wales: Green Building Press, 2nd edition.

Hamilton, Geoff, *Organic Garden Book.* London: Dorling Kindersley, 1997.

Hartmann, Thom, *The Last Hours of Ancient Sunlight.* London: Hodder & Stoughton, 1999.

Hawken, Paul, Lovins, Amory & Lovins, Hunter, *Natural Capitalism: The Next Industrial Revolution.* London: Earthscan Publications Ltd, 1999.

Hillman, Mayer, *How We Can Save the Planet.* London: Penguin Books, 2004.

Hills, Lawrence, *Grow Your Own Fruit and Vegetables.* London: Faber & Faber, 1971.

Howard, Sir Albert, *An Agricultural Testament.* The Other India Press (in association with Earthcare Books), 7th Edition 1956.
Farming & Gardening For Health or Disease. London: Faber & Faber, 1931.
The Waste Products of Agriculture: Their Utilisation as Humus. Oxford: Oxford University Press.

Howard, Louise, *The Earth's Green Carpet.* London: Faber & Faber.

Jeavons, John, *How to Grow More Vegetables, Fruits, Nuts, Berries, Grains and Other Crops Than You Ever Thought Possible on Less Land Than You Can Imagine.* Berkeley, CA: Ten Speed Press, first published 1974, now 5th edition.

King, F. H. *Farmers of Forty Centuries - Permanent Agriculture in China, Korea and Japan.* Now republished with the sub-title: *Organic Farming in China, Korea and Japan.* Mineola N.Y.: Dover Publications, first published 1911, reprinted 2004.

Kuhn, Thomas, *The Structure of Scientific Revolutions.* London: The University of Chicago Press Ltd, 3rd ed.1996.

Lampkin, Nicolas, *Organic Farming.* Tonbridge: Farming Press, 1999.

Lovelock, James, *Gaia: A new look at life on Earth.* Oxford: Oxford University Press, 1979.

Malik, Kenan, *Man, Beast and Zombie.* London: Weidenfeld & Nicolson, 2000.

McCarrison, Robert, *Nutrition & Health.* London: Faber & Faber, 1961.

Merchant, Carolyn, *The Death of Nature.* New York: Harper & Row, 1980.

Mollison, Bill, *Permaculture - A Practical Guide for a Sustainable Future.* Covelo, California: Island Press, 1990.

O'Neill, Amanda, *The Art of Chinese Watercolours.* Siena.

Orwin, Christabel, *The History of British Agriculture 1846-1914.* New edition 1971.

Pfeiffer, Ehrenfried, *Soil Fertility - Renewal & Preservation.* Sussex, England: The Lanthorn Press, latest Edition 1983.

Pretty, Jules, *Regenerating Agriculture.* London: Earthscan, 1995.
Agri-Culture. London: Earthscan Publications Ltd, 2002.

Rodale, Jerome, *Pay Dirt.* New York: Devin-Adair, 1949.

Russell, E John, *Soil Conditions & Plant Growth.* London: Longmans, 11th ed. 1988.

Scheer, Hermann, *A Solar Manifesto.* London: James & James/Earthscan, 2001.
The Solar Economy. London: James & James/Earthscan, 2002.

Schumacher, E.F., *Small is Beautiful.* London: Blond & Briggs Ltd, 1973.

Shiva, Vandana, *Biopiracy - The Plunder of Nature and Knowledge.* Totnes: Green Books, 1998.

Soper, John, *Biodynamic Gardening*. Stourbridge England: Bio-Dynamic Agricultural Association, 1983.

Steiner, Rudolf, *Agriculture*. Stourbridge England: Bio-Dynamic Agricultural Association, 1997.
Spiritual Foundations for the Renewal of Agriculture. Stourbridge: Bio-Dynamic Agricultural Association, 1993.

Sykes, Friend, *Humus & the Farmer*. London: Faber & Faber, 1946.
Food, Farming & the Future. Emmaus, Pa.: Rodale Press, 1951.

Temple, Robert, *The Genius of China*. Prion, 1986.

Tompkins, Peter & Bird, Christopher. *The Secret Life of Plants*. London: Allen Lane/Penguin, 1974.

Turner, Newman, *Fertility Farming*. London: Faber & Faber, 1956. *Herdsmanship*.

London: Faber & Faber. *Fertility Pastures*. London: Faber & Faber.

Voisin, Andre, *Soil, Grass & Cancer*. New York: Philosophical Library, Inc., 1959.

Wookey, Barry. *Rushall: The Story of an Organic Farm*. Oxford: Basil Blackwell, 1987.

PERIODICALS

Earthmatters Friends of the Earth, 26-28 Underwood Street, London N1 7JQ.

HDRA Newsletter Henry Doubleday Association, National Centre for Organic Gardening, Ryton-on-Dunsmore, Coventry CV8 3LG

New Scientist King's Reach Tower, Stamford Street, London SE1 9LS

The Living Earth Soil Association, 86 Colston Street, Bristol BS1 5BB

Star & Furrow Biodynamic Association, Orchard Leigh Community, Bath Road, Eastington, Stonehouse, Gloucestershire GL10 3AY

PUBLICATIONS & REPORTS

Soil Association, 1996. *Counting the Costs of Industrial Farming*.

Soil Association. *Organic Farming, Food Quality & Human Health*.

The Composting Association, 2001. *The State of Composting 1999*.

The International Institute of Biological Husbandry, 1978. Review Paper Series, No.1. *Who Needs Inorganic Fertilizers Anyway?* by R.D. Hodges Ph.D, Wye College, Ashford, Kent.

Useful Organisations and Websites

American Chemical Society
www.acs.org

Bio-Dynamic Agricultural Association
Woodman Lane, Clent
Stourbridge, West Midlands
DY9 9PX
www.biodynamic.org.uk

British Organic Farmers
86 Colston Street, Bristol BS1 5BB

The Composting Association
www.compost.org.uk

Composting Trends and Technologies www.cals.cornell.edu/dept/compost.trends.html

Department for Environment, Food & Rural Affairs
www.defra.gov.uk

Druker. Steven M.
sdruker@biointegrity.org

Elm Farm Research Centre
Hamstead Marshall
Newbury, Berkshire RG15 0HR
www.efrc.com

Forestry Stewardship Council
info@fsc-uk.org

Friends of the Earth
www.foe.co.uk

Henry Doubleday Research Association
National Centre for Organic Gardening
Ryton-on-Dunsmore, Coventry CV8 3LG
www.hdra.org.uk

Dr Ingham
www.soilfoodweb.com

Institute of Grassland & Environmental Research
Plas Gogerddan
Aberystwyth SY23 3EB
www.iger.bbsrc.ac.uk

Irish Organic Farmers & Growers Association
14 Berkley Road, Dublin 7, Eire

New Scientist Magazine
www.newscientist.com

Organic Growers Association
86 Colston Street
Bristol BS1 5BB

Organic Worldwide News
www.organicts.com

Positive World-wide News
www.global-good-news.com
(agriculture)

The Rodale Institute
www.NewFarm.org

Soil Association
86 Colston Street
Bristol BS1 5BB
www.soilassociation.org

World Food Program
www.wfp.org

Index

Aberystwyth Agricultural College, University College of Wales 72, 194
actinomycetes 17
Action Plan to Develop Organic Food and Farming in England (UK government initiative) 218-9
atmosphere, self-regulation 16
Advisory Committee on Agricultural Feeds, UK 123
Advisory Committee on Releases to the Environment (ACRE) 123
Africa 24, 207
 pest control in 81
 organic farming in 226-7
Africa, North 165, 211
AgBioWorld.org 237
Age of Enlightenment, The 169-70, 158, 162, 163, 172, 178
Age of Reason 45
Agent Orange 180
agricultural best practice 13-14, 46
 colleges and farmers 59
 chemistry 9
 and environment 203
 equipment, development of 175
 industry: training 193-4
 vested interests 89
 policies 227
 planning 230
 scientists 212
 subsidies 102
 systems, sustainability 181
 in transition 236-8
agricultural efficiency and productivity 190-3
Agricultural Product Disparagement Laws, US 122
Agricultural Renewal in India for Sustainable Environment (ARISE) 215
Agricultural Research Station, Pusa, Bengal 50
Agricultural Revolution 42, 171, 175

Agricultural Testament, An (Howard) 10, 52, 56, 57, 67, 111
Agri-Culture (Pretty) 152
agriculture, definition of 13-14
 conventional, and pesticides 98
 traditional methods of 35
 Western, future for 184ff
agro-chemicals 204, 231
agro-companies, power of 31
agro-ecological movement, Africa 226
agroforesty 226
alcohol as fuel 187
algae, soil 71
 paddy fields 206
Alliance for Bio-Integrity 125, 128
Allobophora caliginosa 72
allotments 136
Amazon, South America, soils 233
ammonia 16, 66, 67, 133, 135, 146, 151
Animal By-Products (Amendment) Order 2001, The 141, 142
animal feed 30, 228
animal husbandry, trends in 33, 84
 manure 136
 welfare 204
animals
 farm, and antibiotics 103
 healthy 79
 respect for 20-21, 32
 and rotation, on farms 47, 84
antibiotics 89, 203, 211, 221
 use of 81
 residues 102
antioxidants and organic fruit 90-1
ants as pest control 228
aphids 85
apples 91, 94, 96
 organic 218
Aquinas, Thomas 165
Argentina 27, 147
 organic farming in 231
ARISE *see* Agricultural

Renewal in India for Sustainable Environment
Aristotle and Aristotelian view 161-2, 165, 167
artificial hormones 222
Asia 23, 207, 209, 211
 rice farming 200
Augustine, St 166, 168
Australia 197, 198
Austria 142, 216
Aventis 117, 123
Avery, Denis T. 91-92, 237
azola/azolla fern 200, 206-7, 223
Azotobacter bacteria 134
baby food, non-organic, survey of 96
Bacillus thuringiensis 79, 80, 108, 114, 117, 119, 228
bacillus 80
Bacon, Francis 167-8, 170
bacteria 14
 biological control 80
 and composting 134
 and methane production 151
 see also soil bacteria
Balfour, Lady Eve 10, 34, 47, 54-56
bananas 50, 104, 212
 and GM regulations 209
barley 40, 147, 165, 195
beans 77, 84, 191, 201, 228, 234
 as green manure 146-7
beef production 210-11
Benbrook, Dr Charles 117
Bennett, Stanley 222
benzene hexachloride (BHC) 99-100
benzene 100
BHC *see* Benzene hexachloride
BigBarn.co.uk 190
biodigesters 154
biodiverse farms 205
biodiversity 14, 18, 32, 77, 79, 102, 191-2, 201, 208, 227
biodynamic methods in farming 195
 system of organic farming 53-54

Index

Biodynamic Movement (Steiner) 35
bioethanol 189, 190
biofeedback 16-17
biogas plants and production 152, 185-6
'biointensive' gardening and farming 201
biological control 79, 80-81
biological sciences 10
biomass, power production from 187
biopesticides 228
Biopiracy (Shiva) 191-2, 235
biosphere 15
biotechnology 91-92, 237
Biotechnology Industry Association, US 121-2
biowaste 143
birds, effects of pesticides on 101
Blake, William 176
blue-green algae 150
Bohm, David 163
bollworm 114, 115
bones, powdered 174
Borlaug, Norman 24
bovine spongiform encephalopathy (BSE) 87, 13, 102, 104-106, 120
box schemes for food 189
brandling worms or tiger worms (*Eisenia foetida*) 73, 134
Brazil 147, 187, 230, 231-3
break crops, use of 46
British agriculture, review of 192
British Association of Public Analysts 94-95
British Medical Association (BMA): attitude toward GM foods 118
British Trust for Ornithology 101
Bromfield, Louis 58-9
Brown, Linda 12, 14, 99, 208
BSE *see* bovine spongiform encephalopathy
budworm 114
bullfinch 101
bunting, reed 101
butterflies, Monarch 117
cabbage white butterfly, and cabbage moth caterpillars 79, 228
cadmium 149, 151

California Certified Organic Farmers (CCOF) 223
California 223
Cambridge University Agricultural Economics Association 195
campylobacter 103
Canada 116, 160
  agricultural exports 184
  GM crops 113, 126
cancer and pesticides 97, 98
cane sugar and energy production 187
canola, GM 127
Capra, Fritjof 162, 191, 192-3, 236
carbon dioxide 19, 20, 67, 143, 152, 153, 155, 172
carcinogenic compounds 94, 98, 104, 109
Caribbean farmers 209
carrot fly 81
carrots 88-89, 95
Carson, Rachel 35
cashew farming, organic 224, 225
cattle and biogas production 185
cattle farming 60, 199
  and BSE crisis 104-6
  GM maize and 147-8
cattle 51, 217
CCOF *see* California Certified Organic Farmers
CDC *see* Centers for Disease Control and Prevention
Cecap *see* Central Cordillera Agricultural Programme
cellulose 19, 66, 67, 70, 92, 133, 134, 137, 146, 151, 154, 186, 200
Centers for Disease Control and Prevention (CDC), US 93
Central Cordillera Agricultural Programme (Cecap) 206-7
Central Institute for Cotton Research (CICR) 115
Central Science Laboratory 88
Centre for Environment and Society, Essex University 102
cereal agriculture and crops 43, 85, 102, 212
Chapela, Ignacio 115, 116, 118

charcoal and soil improvement 233
Charles, Prince of Wales 107, 153
cheese making co-operatives, France 198
chemical toxicity 213
Chiapas state Mexico 230
  farmers of 191
chickens 21, 103, 106, 117-8, 223
chicuna beans 234
China 39, 42, 152, 174, 170, 182, 209, 223-4
  agriculture 10, 34, 37, 38, 47, 51, 84
  and GM crops 113
  and Marxism 179
Chinese clover 40
Chinese organic farming 9
chlorinated hydrocarbons 149, 151
Christianity and attitudes towards nature 164, 165
CJD *see* Creutzfeldt-Jakob Disease
climate change 182
clover 18, 19, 20, 42, 43, 46, 69, 77, 147
'cocktail effect' 93, 98-99
coconut farming, organic 224
coffee, organic 225
  production 230
commodity prices, movements in 25
common land 175
Community Forests 203
community land management 230
companion planting 201
comparative yields, organic and conventional 194-6
complementary approaches, for livestock 81
compost as a pollutant control agent 145-6
compost 54, 78, 132, 141, 150, 200, 228
  and mulching 229-30
  production of 51
  separation of components 142
compost, liquid extracts of (compost teas) 80
Composting Association, The 143-4
composting methods, large

scale (window system, aerated static pile, in-vessel systems, fully contained rotating system) 139-140
composting sites, placement of 140-1
composting 39-40, 41,
  carbon and 133-34
  community 136
  farm 137
  household and gardening 135-36
  municipal 134 137-38
  in New York State 138
  theory of 133-35
  in UK 136, 138, 140
Conford, Philip 35, 36
Consultative Group on International Agricultural Research 24
consumers and organic farming 217
conventional farming 10, 12-3, 22-31, 87
Co-operative Oilseeds Growers' Federation 224
co-operatives 198
coppicing 202
  coppicing woodland 187
corn, GM 127
cosmic rhythms 53
cotton production, India 215
cotton 40, 41, 212, GM 108, 114-5, 213, 214
  farming subsidies, US 196
  organic 215
credit schemes, farming 212
Creutzfeldt-Jakob Disease (CJD) 104
crop nutrition 30, 172
crop residues 86, 148
crop rotation 42-43, 44, 46-7, 60, 77, 83, 86, 228, 229
crop yields 205, decline in 213, 214
crops, complementary 83
crops, traditional, patenting of 129-30
Cuba 221
  farming in 77, 228-9, 236
Curry, Sir Donald 192
Czech Republic 216
Daily Express 98
Daily Mail 122
Darwin, Charles 177-8, 179
Darwinian natural selection 16
Davy, Humphrey 172

DDT 56, 94, 99, 104
de Orellana, Francisco 233
De rerum natura (Lucretius) 164
Dean, Baroness 89
debt, rural 213
decontaminants 145
DEFRA *see* Department of the Environment, Food and Rural Affairs
'Demeter' range of products 54
Denmark 142, 194, 216, 230
Department for Trade and Industry, UK 188
Department of Agriculture, Barbados 50
Department of Agriculture, US 117
Department of Energy, US 182
Department of Human Anatomy and Cell Biology, Liverpool University 99
Department of the Environment, Food and Rural Affairs (DEFRA), UK 89, 103, 181, 188, 89, 218
Derris 79
Descartes, René 169-70
  Cartesian paradigm 171
desertification 226
dieldrin 99
diesel, synthetic, production from sewage 151
diet, national, and effect on longevity 48
  and health 54-5
digesters, anaerobic 151, 152
Dimbleby, Jonathan 237
Dinoseb 98
disease control 141-2
diseases 82, 106, 203
diversity, loss of 81
  and agricultural production 205
DNA 113, 114, and GMO 118
Dow Chemicals 180
Drinking Water Inspectorate, UK 100
Drinking Water Standard, EU 100
drug residues 203
Druker, Steven 125-26, 128-9
dustbowl, US 179-80

E. coli 92, 103
earth science 72
earthworms (*Lumbricus terrestris*), and effects on soils 17, 63, 72-74, 82, 102
Eastern perceptions of Nature 82, 162-3, 179
EC Working Document: Biological Treatment of Biowaste 143
enchytraeid worms 17
Eclipse Scientific Group laboratory, Cambridgeshire 88
ecological sciences 10
ecological approaches 198
ecological practices, Switzerland 219-20
ecological studies 194
ecologists 81
  and perceptions of humanity 159
ecology 15, 56, 68
economics
  and agriculture 190-1
  'Mickey-Mouse economics' 196
economists, and measuring efficiency of business 190-1
economy and rural development 204
ecosystem 18, 76, 81, 203, 209
Ecuador 207
efficiency, agricultural 191-3
EFTA (European Free Trade Area) 219
Egypt 161, 226
Egyptians, Ancient 80
Elm Farm Research Centre 55, 62, 188
Enclosure Acts 43
enclosure movement 43-46
energy Crisis 181
Energy Crops Scheme, DEFRA 188
energy crops 187-8
English Nature 116-117
environment 191, 203 ff
Environmental Agency, US 151
environmental pollution 100
Environmental Protection Agency (EPA), US 98
EPA *see* Environmental Protection Agency

Index

epidemiological studies 99
erosion 27
Ethiopia 226, 230
EU *see* European Union
Europe 147, 169, 236
  and GM foods 120, 123
  and composting 138
European attitudes toward nature, historical 25-26, 34 ff
European Consortium for Organic Plant Breeding (ECO-PB) 55
European Organization for Renewable Energies 182
European Union (EU) 141, 209, 225-6, 231
  agricultural comparisons 192
  Drinking Water Standard 100
  and farming 215-6
  farm subsidies 196
  and GM crops 125, 127, 128
  Land Directive 143
  and landfill 132
  regulations on pesticides 99
  support for farm reform 197
  training programs 194
EUROSOLAR 182
Evans, Bob 27
factory farming 91, 237
Fagan, Dr John 109-10
fair trade movement 208-9
FAO *see* Food and Agriculture Organisation
farm animals and GM crops 127
farm shops 189
Farm Woodland Premium Scheme 203
Farmer Field Schools (FFS) 227
farmers
  and supermarkets 189
  Third World 25
  Western 25
Farmers for Forty Centuries: Permanent Agriculture in China, Korea and Japan (Organic Farming in China, Korea and Japan) (King) 10, 13-14, 37, 49, 149, 200
'Farming for Profit with Organic Manures as the Sole Medium of Refertilization' (Sykes) 57
farming
  comparing conventional and organic 19
  communities 204
  definition of 13-14
  English 17-18
  high yield 91
  in transition 236-8
  in the Third World 31
  in the West 31, 33
  industrial 102
  methods 34, 204
  mixed 77, 199
  modern 23, 255
  oriental 47, 51, 62
  see also King, F. H.
  practices, classical, unsustainable 165
  practices in Switzerland 219-20
  systems 83, 205
  developments in 210
  Third World
  traditional 205
  unsustainable methods of 179
farmland wildlife 203
farms 17-19
  and composting 140-1
  non-organic 26-27
  organic and other, comparisons 62-3
Faulkner, Edward 10, 59
FDA *see* Food and Drug Administration
feldspar 54
fertilizers 28, 29, 30, 51, 56, 63, 85, 101-102, 152, 173, 181, 183, 191, 207, 212-3, 226, 227, 228, 232, 233
  artificial 221
  chemical 17-8, 26
  cost of 27
  development of 174
  liquid 38
Fertility Farming (Turner) 10, 59
fertility of the soil, natural 26
Fertility Pastures (Turner) 59
FiBL *see* Research Institute of Organic Farming
Finland 216
fish farming, China 38
  and rice paddies 223
fish-meal 105
Flowerdew, Bob 12
Food and Agriculture Organization (FAO) 205
Food and Drug Administration (FDA), USA 121, 128, 222
  and fraud 125-6
food and other allergies 203
food chain 204
food commodity prices 193
food industry and environment 203
food miles 189-90, 204
food production 204,
  growth in 23
food productivity in crisis 195
food quality 87ff
Food Standard Agency (FSA) 87-89, 103, 118
Food Standards Authority 237
food waste and compost 138, 141, 143, 145, 154
Food, Farming and the Future (Sykes) 56
food-derived illnesses 118-9
foot and mouth disease 51, 141
forest farming 199
forest gardens 207-8
Forestry Commission 202-3
forestry 151
  sustainable 202-3
fossil fuels 181, 182, 184
France 216-7, 230
  agricultural development 42
'Frankenstein foods' 107
free trade 91, 237
Friends of the Earth 237
From my Experience (Howard) 58
fruit 18, 48, 50, 191, 199, 200, 221, 230
  British tests on 94-5
  mineral and nutrient levels in 90-1, 202
  organic 230
  pesticide levels in 93-5, 96-7
  zero-tillage and 147
FSA *see* Food Standard Agency
fuel cells (methane-fired) 152
fuel crisis 182, 184, 188
FuelCell Energy Inc 152
Fukuoka, Masanobu 10, 147, 148

fully contained rotating systems (composting) 139
fungal diseases 50
 control 80, 84
fungal mycorrhizae 29
fungicides 23, 28, 80, 102
Gaia (Mother Earth) 15-17
Gandhi, Mahatma 214
Garden of Eden 160, 169
garlic 80
gene regulation 109
'General Principles of Food Law in the EU, 30 April 1997' (EU Commission) 127
genetic engineering 109-110
genetically modified (GM) foodstuffs 13, 14, 24, 107, 125
 GM plants and politics 121 ff
 reseeding of by farmers etc 122, 124
 GM crops 31
 and genetic variation 127-28, 131
 GM products, health implications 117-119
 GM research 108
 GM technology 237
 GM genes 109, 113-5, 118, 120
 GM maize 115-6
 GM foods 107 ff
genetically modified organisms (GMOs) 107 ff, 119, 221
Germany 80, 180, 216, 230
 energy production 187
Ghana Organic Agricultural Network (GOAN) 227
Ghana 226
global warming 33, 181, 187, 227
GM companies and politics 121 ff
GM crops and insects 108
'GM Nation?' (public debate, UK) 123, 131
GM seed retention 120, 122, 124
GM see genetically modified heading
GMOs see Genetically Modified Organisms
Golden Rice 112, 113
government and recycling 140

grasses 77
Greeks, ancient 161, 163, 165
green manure 39-40, 41, 78, 86, 132, 186, 201, 202, 234
green manuring 146-147, 148, 228
Green Party, US 222
Green Revolution 35, 111, 170-1, 173, 209, 211-14
 systems 205
Greenpeace 237
Grey, Lord (chairman of ACRE) 123
groundnut 50
Grow Your Own Fruit and Vegetables (Hills) 10, 12
growth hormones 211, 221
Grupo de Agricultura Organicas (GAO), Cuba 228
Guano, Peruvian 174
Guatemala 200, 202, 221, 230, 234
Gulf of Mexico, pollution in 210
hardwood trees 202
Hardy, Sarah 89
Hartmann, Thom 159-60
Harvest Help, UK charity 227
Haryana region, India 170, 211, 213
Haughley, Suffolk, UK 54, 55
HDRA see Henry Doubleday Research Association
heavy metals 138, 142, 145, 148-9, 150-1, 153
 soil contamination 145
hemp 41
Henry Doubleday Research Association (HDRA) 61, 227
herbal remedies 81
herbicides 23, 100, 101, 117
herbivores and animal protein feed 105
herbs 78
Herdsmanship (Turner) 59
hexachlorobenzene 99
Highgrove Estate, UK 153
Hillman, Mayer 181-2
Hills, Lawrence D. 10, 12, 61
Hobart, Tom 121-122
holistic approach 52, 53, 82

holistic view of agriculture 51-52
Holland, agricultural development 42
homeopathic medicines 53, 81
Homer 161, 164
Honduras 200, 202, 230, 234
Hormone Disclosure Law, Vermont 222
horticulturists, organic 80
household waste separation 142
housing estates and sewage treatment 153
How Can We Save the Planet (Hillman) 181-2
How to Grow More Vegetables, Fruits, Nuts, Berries, Grains and Other Crops Than You Ever Thought Possible on Less Land Than You Can Imagine (Jeavons) 201
Howard, Sir Albert 10, 11, 34, 36, 47, 49-52, 54, 55, 56, 57, 58, 59, 61, 62, 63, 65, 67, 70, 77-78, 83, 111, 112, 132
 and composting 133, 134, 199
human health and GM foods 117-9
human sewage recycling 39
 see also sewage
humus 27-28, 29, 52, 55, 62, 66-76, 82, 148, 135, 137, 172, 179, 185, 199, 200, 232, 234
Hungary 216
hunter-gatherer lifestyle 159, 161
Hunzas, Indian group 48, 49
hydrogen as fuel 186-7
IFOAM see The International Federation of Organic Agricultural Movements
incinerators 140
India 152, 182, 209, 211, 213, 221, 224-6, 230, 236
 agriculture 47, 50-51
 agricultural changes in 170-1
 dietary comparisons in 48
 farming 34, 201, 211, 213
 and GM plants 114-5, 120, 121, 124

Monsanto in 114-115
rice farming 200
Indian communities, Brazil 232
'indigenous best practices' (Cecap) 206
Indonesia 205, 207, 236
'Indore Process' 51
industrial revolution 156-7, 175, 178
insecticides 23, 56, 79
Institute of Ecology, Mexico 116
Institute of Grassland and Environmental Research, Aberystwyth 91
Institute of Plant Industry, Indore, India 51, 58
integrated biodiversity 17-19
integrated pest management (IPM) schemes 228
intensive farming 19
intensive animal rearing 30, 212
intensive organic urban horticultural programmes, Cuba 229
intensive planting 201
intercropping 228
International Conference on Organic Products 224
International Federation of Organic Agricultural Movements, The (IFOAM) 21-2, 215
International Fund for Agricultural Development (IFAD) 35
International Monetary Fund (IMF) 35
International Research Association for Organic Food Quality and Health 55
intertillage 40-41, 77, 84
in-vessel composting 141
ioxynil 98
IPU (herbicide) 100
Ireland 188
Irish potato famine 173
iron content of apples 91
irradiation 91, 221, 237
irrigation schemes 212
irrigation systems 213, 230
Italy 194, 215-6, 230
deforestation in Classical period 165
Japan 34, 39, 40, 221, 230, 231

and GM products 128
and hydrogen fuel cell vehicles 186-7
agriculture 37
Japanese organic farming 9
agriculture 10
Jatan ('care') movement, India 214-5
Java 205
Jeavons, John 201
John Innes Centre, UK 128
Judaism 160, 161
Judeo-Christian tradition 164
Kansas, cattle farming in 210-11
Kant, Immanuel 175-6
Kenya 110, 207, 221, 230
original agricultural practices in 36
permaculture and 199
Kenyan Agricultural Research Institute 110
kerbside collections, UK 144
King, F. H. 10, 11, 37-42 47, 49, 149, 200
Kite's Nest, Worcestershire, cattle farming in 211
Korean agriculture 37
organic farming in 9, 34
Korean organic farming 9
agriculture 10
Krebs, Sir John 87-88, 89, 91, 237
lamb 189, 195
Lampkin, Nicolas 9, 10, 12, 57, 84, 92, 97, 104, 183, 194, 195, 236
Lancet, The 118
Land Directive, EU 143
land management, ecological 207-8
land reclamation 144
landfill 132, 140, 142, 143, 150
sites 144
lapwing 101
Last Hours of Ancient Sunlight, The (Hartmann) 159-60
laterite soils, and mineral deficiencies 232
Latin America 207, 209, 211, 229-34
'Law of Return' 10, 19-20, 32
Law of the Minimum 172
Lawes' superphosphate 174

lead 149, 150, 151
legumes 202
organic 230
leguminous crops, and nitrogen fixing 40
Liebig, Justus von 12, 13, 22, 25, 74, 172
lignin 66, 70, 134
lime 56-7
Lincolnshire, farming in 17
Lindane 94, 99
livestock and pest control 81
livestock farming 212
Living Soil, The (Balfour) 10, 55
Living Soil and the Haughley Experiment, The (Balfour) 55
London Food Commission 98
Lovelock, James 15-16
lucerne 60
Luddism, accusations of 108, 112, 121, 122
Lumbricus terrestris (common earthworm) 72
lupins 188
Mader, Paul 62-63
Mahco (Monsanto-Mahyco) 115, 120
maize 30, 40, 77, 234
GM and non-GM 115-6, 127
organic 230
in US 210
Malabar Farm, US (also book of that name) 58
Malawi 227
Malaysia 152, 223
Malik, Kenan 158-9, 167, 173
Man, Beast and Zombie (Malik) 158-9, 167
'Manufacture of Humus by the Indore Method, The' (Howard) 52
manure 19, 20, 26, 46, 51, 53, 63, 92, 133, 136, 146, 147, 148, 212, 222
organic 225
worms 136
Marx, Karl 159
Marxism 179
Massachusetts Institute of Technology 128
Maxwell, Phillip 114
Mayall, Sam 72
McCarrison, Dr Robert 36, 47, 48-49, 55

McNaughton, David 219
Meat & Livestock
  Commission, UK 103
meat 23, 136, 198, 211
  antibiotic residues in 91
  BSE and 105
  GM and 127
  organic and conventional
    compared 91
  imports to UK 189
  see also under individual
    meats
meat-processing
  co-operatives, France 198
mechanical pest controls 81
Medicago astragalus 40
Medieval European Vision
  165-6
medieval farming,
  insufficiencies of 44-5
medieval system of
  agriculture 43-44
Mesopotamia 161
methane Production 151-3
methane 16, 151-3, 154
Mexican Coffee Council 230
Mexico 24, 115-116, 182,
  229-30
microlife 53
micronutrition 213
Middle Ages 156
mildew, powdery 84
milk, organic and
  conventional 91-2
  and residues 99-100
  production 195
millet 40-41
minerals, use in pest control
  80
Minimum Residue Limits
  (MRL) 93, 95
Ministry of Agriculture, India
  115
Ministry of Agriculture,
  Netherlands 216
Ministry of Agriculture, UK
  87, 91
Ministry of the Environment,
  Mexico 116
Minoan civilization 163
Mitchell, Dr Alyson 90
mites, predatory
  (*Phytoseiulus persimilis*)
  80
molacus beans 234
Mollison, Dr Bill 198
monocrops 111, 152, 213,
  228

monoculture 173, 192, 212
Monsanto 108, 114, 115,
  117, 120, 121, 124, 127-8,
  180, 222
MORI poll 217
Mother Earth (Gaia) 15-17
Mother Nature 173
MRL Minimum Residue
  Limits
mucuna mean and soil
  improvement 234
multiple cropping 40-41
municipal compost, uses of
  144-6
municipal solid waste 137-8
Mutual Aid (Kropotkin) 178
mycorrhizae fungi and
  threads (hyphae) 62, 63,
  28, 68-71, 76, 81, 85, 232
Nader, Ralph 222
napalm 180
National Farmers Union, US
  124
National Farmers' Union, UK
  193-4
National Federation of Young
  Farmers' Clubs 194
National Forest Company
  203
National Grid 187
National Institute for
  Agricultural Botany 84
National Pollutant Discharge
  Elimination System
  Regulations, US 145, 151
Native American agriculture
  35, 36
natural philosophy 173
nature reserve areas 78
Nature 11, 13, 14, 17, 20,
  25, 32, 33, 47, 56, 57, 61,
  65, 76-77, 78, 82, 86, 106,
  132, 133, 161-2, 164, 165,
  166, 167, 169, 173, 174,
  175, 176, 180, 192, 235,
  238
  definition of 157
  Western attitudes towards
    25-26
  Western perceptions of 155
    ff
neem tree 80, 212
Nelson, New Zealand 151
Neo-Synthesis Research
  Centre (NSRC), Sri Lanka
  208
Netherlands 142
  agriculture in 37

New Farm Magazine, The
  221
NewFarm.org 221
New Zealand 197, 221
Newton, Sir Isaac 34,
  169-70
Nicaragua 230
Nigeria 205
nitrates 26, 102, 104
nitrogen 132, 146, 152, 172,
  214
  and compost 51
  and plants 74-75, 206
  fertilisers 149, 180, 210,
    212
  fixing of 71, 135, 150
nitrogen and soil 53-4
nitrogen, phosphorus and
  potash (NPK) 174, 72,
  213
NOP poll 217
no-plough technique 59, 85
'Norfolk four-course rotation
  system' 17-18, 43, 202
Norfolk, farming in 17
Normandy, France 198
Northwest Science and
  Environmental Policy
  Center, Sandpoint, Idaho
  117
no-tillage techniques *see*
  zero-tillage
Novartis 116
Novum Organum (Bacon)
  168
'NPK' method 62, 145, 172
  see also nitrogen,
    phosphorus and potash
nutrients, banked and
  invested 74-76
Nutrition and Health
  (McCarrison) 49
Nutritional Research
  Department, India 48
nuts, organic 230
oats 43, 57, 84, 195
OEE (Office of Special
  Studies), Mexico 24
Office of Science and
  Technology, Department of
  Trade and Industry, UK
  122
Office of Special Studies
  (OEE), Mexico 24
oil supplies, future decline in
  33, 182
Omega 3 essential fatty acids
  91

Index

One Straw Revolution
  (Fukuoka) 10, 147
Orellana, Francisco de 233
Organic Advisory Service 55
organic agriculture 87
  courses on 193-4
  effects of pollution on 104
Organic Bible, Successful
  Gardening the Natural
  Way (Flowerdew) 12
organic exports 208-9
Organic Farming (Lampkin)
  9, 12, 57, 84, 183
organic farming 24, 92, 106,
  179, 181, 221, 208-9, 220,
  237-8
  Argentina 231
  Britain 55
  Chinese, Japanese, Korean 9
  and GM 130
  definition of 14-17
  Gujarati meeting on 215,
    UK 217-8
  history 9
  in practice 17-22
  India 214, 224, 225-6
  modern, influences on 46
  pioneers 42
  plant diseases, without
    animals 199-202
  'Organic Farming, Food
    Quality and Human
    Health' (Soil Association
    report) 88, 92
organic farms 27
organic food
  China 223
  and flavour 91
  imports to UK 190
  negative presentation 92
  purchases of 88, 189
  scientific studies 90
  see also under individual
    foodstuffs
Organic Gardening and
  Farming magazine 58
organic growth 236
organic husbandry 12, 234
  and pollutants 33
organic movement, growth
  of 34 ff
  and GM crops 130
organic philosophy 21-22
organic pioneers 42, 47-8,
  234-5
Organic Price Index 221
organic products, sales in US
  222

organic recycling 132
'organic revolution' 234-8
Organic Trade Association
  (OTA), Massachusetts 220
organic waste 19, 20, 78, 132
Organisation for Economic
  Co-operation and
  Development (OECD) 197
organochlorines 100
organophosphates 100
Oriental culture 156
Origins of the Organic
  Movement (Conford) 35
Out of the Earth (Bromfield)
  58
overproduction of food 196-8
Pacific Rim markets 197
Paraguay 147
Patagonian lamb 231
patenting of traditional crops
  129-30
paternalism 172-3
PCBs *see* polychlorinated
  biphenyls
peas 83, 84
peat, eliminating use of 144
pelton wheels 186
penicillium 80
perceptions of history, linear
  and cyclical 179
Permaculture One (Mollison
  & Holmgren) 198
permaculture 198-9
Peru 230
pest control 84, 226,
  mechanical 81
pesticide residues 92, 203
  and British Association of
    Public Analysts survey 94
  and FAA 88
Pesticide Residues
  Committee, UK 93, 96-97
Pesticide Safety Directorate
  93
pesticides 14, 30, 51, 63, 79,
  84, 89, 91-92 102, 106,
  180, 191, 212, 213, 226,
  227, 228, 237
  and foodstuffs 92-97
  banned, use in the Third
    World 100
  cost of 27
  costs of removal from water
    101, 102
  in drinking water 100
  and GM crops 114, 117,
    119
  and GMOs 117

harmful effects of 97-106
  in Switzerland 219-20
  and tests on animals 93-94
  resistance to 179
pests 82, 83, 106
petrochemical fertilizers 237
Philippines 207, rice farming
  200
phosphorus/phosphates
  and plants 74, 76
  and composting 135
photovoltaic (PV) panels
  186, 187
pigs 18, 103, 195, 199
  and biogas production 185
pink clover (*Astragalus*
  sinicus) 40
plant diseases, susceptibility,
  variations in 84-85
plant extracts, as pest control
  80
plant nutrition 19, 22, 171,
  174
planting
  intensive, companion 201
plants, healthy 79
Pleasant Valley 58, 59
Ploughman's Folly 10, 59
plums, organic 218
  pesticides in 194
Plunkett, Donald 24
Policy Commission on the
  Future of Farming and
  Food (DEFRA) 181, 217
pollutants 204
pollution 29-30, 100-106,
  191, 196, 203, 208
  effects on organic husbandry
    104
  IFOAM and 21
polychlorinated biphenyls
  (PCBs) 99
ponds 78
population growth and
  farming 183-4
Population Institute,
  Washington 184
pork 223
Portugal 199
potash/potassium and plants
  57, 74, 75-76, 172
  and compost 135
potato blight 79
potatoes 83, 218
  comparison of organic and
    conventional 91
  pesticides in 96, 97, GM
    118

Prakriti (Prakruti) 158
Pretty, Jules 12, 36, 87, 102, 106, 112, 152, 171, 193, 205, 206-7
Prevention magazine 58
prions 105
production, highly intensive 37
Protagoras 166-7
Public Health Association of Australia (PHAA) 127-28
pumpkins 181
Punjab region, India 170, 211, 213
Pusa 50-51
PV *see* photovoltaic panels
pyrethrum 79
quartz 54
quassia 80
Quist, David 115, 116
rainforest 232
 destruction 231
rainwater 148
 recycling 38
rape plant 38, 46
rape seed 130
RBST hormones 222
recycling 10-11, 13, 19, 38-39, 78
red clover 201
red spider mite 80
redwing 101
reed bed technology 153-4
Regenerating Agriculture (Pretty) 36, 171
regional food, UK 189-90
Regulation on Animal By-Products (EU) 141, 142
Renaissance, The 163, 166-7
renewable energy 185
Repetto, Robert 192
Research Institute of Organic Agriculture 62
Research Institute of Organic Farming (FiBL) Switzerland 225
research stations 51, 207
Rhizobia bacteria 20, 42, 68-9, 74
Rhizobia-containing crops 20
rhizosphere 13, and bacteria 68
rice blast fungus 84
rice farming 84, 147, 200, 211
 Philippines 206
 China 223

Rice Tec 129
rice 97, 147, 206-7, 212
rinderpest 51
'Risk Assessment of Mixtures of Pesticides and Similar Substances' (UK government report) 99
Rodale Institute 220
Rodale, Anthony 220-1
Rodale, Jerome 34, 57-58
Roman Empire 165
Roman view of Nature 164
Romantic Movement 175-7
Romantic Poets 176
root-rot disease, cotton 115
Rothamsted Experimental Agricultural Station 62, 150
Roundup weedkiller 108
Roundup-ready soybeans 117, 124, 128, 130
Rousseau, Jean Jacques 168-9, 175-6
Rowett Institute, UK 118
Royal College of Science, UK 50
Royal Society for the Protection of Birds (RSPB) 29-30, 101
Royal Society of Canada 127
Royal Society of Chemistry 91
Royal Society 118
Rumex vesicarious (tak palang) 113
Rushall farm, Wiltshire 218
Rushall: The Story of an Organic Farm (Wookey) 85
Russell, Sir John 62
russian comfrey 18
rye 147
Saami peoples, Norwegian Arctic 159
Sahara Desert 165
Sainsbury, Lord (Under-Secretary of State for Science, UK) 122
Sajiv kheti farming, India 214
salmonella 103
Scheer, Hermann 182
School of Agriculture, Cambridge University 52
Science (journal) 62-63, 98
science and spirituality 234-5
scientific and other research on GM crops 125-26

scientific insights into agriculture 61-3
Scientific Revolution 170
scientific studies of organic food 90
scientists, attitudes towards 24
scrapie 104
seed varieties, promotion of 226
self-sufficiency 32
'self-sustaining super organism', Earth as (Lovelock) 15-16
septicaemia 51
Serratia 80
sesame, organic 230
sewage treatment, composite methods 154
sewage 148-9, 186
 and organic husbandry 104
sewage-sludge composting 141, 188
Seymour Johnson Air Force Base, North Carolina 145
Shah, Kapil 214, 215
sheep 109, 199, 217, use in crop rotation 43
sheet composting 146-147
Shiva, Vandana 14, 15, 26, 109, 112, 167, 173, 191-2, 205, 215, 235, 237
Shopper's Guide to Organic Food, The (Brown) 12, 14, 99, 208
shrimp farming, organic 223
Shropshire 26, 46
silage 60
 crop, GM 117-118
Silent Spring (Carson) 35
silk 41
sludge (sewage and other) 150
 contamination of, by heavy metals 150-1
sludge, toxic 221-2
slugs 79
soil additives 53
soil animals 71
Soil Association 9, 19, 20, 21, 55, 88, 92, 106, 150, 185, 194, 195, 237
 and Policy Commission on Future of Farming and Food 217
 recommendations 203-4
soil bacteria 67-71, 73, 174, 232, 233

Index

bacterial activity 178, 212
micro-organisms 63, 67-68
organisms and use of sludge 150
Soil Conditions and Plant Growth (Russell) 62
Soil 13, 17, 19, 28-29, 27
  chemistry 11
  conservation 228
  contamination 145
  degradation 27-28
  ecology and depletion 212-3
  ecology 65-66, 82
  ecosystem 13
  erosion 37, 148, 197
  feeding of 19
  fertility 9, 20, 22, 226, 229
  healthy 32, 79
  pollution of 101-102
  science 174
  structure of 78
  testing 23
soil 133
solar energy 187
Solar Manifesto, A (Scheer) 182
solid waste recycling 137-38
Sophocles 61
South Africa 226
South America 147-8, 209, 232, 234
Soya UK (organisation) 219
soya 130, 219, 231
  allergies to 118
soybeans, organic 230
Spain 199, 230
Spices 225
Spices Board of India 224
spirituality 234
squash (vegetable) 109, 201
Sri Lanka 207, 208
Staphylococcus infections, US 103
Steiner, Rudolf 34, 35, 47, 49, 52-4, 200
Stewart, Dr V. I. 72-74
stinging nettle 80
Strategy for Sustainable Farming and Food (UK government initiative) 188, 199, 202-3, 218, 219
straw 10, 18, 212
  as animal bedding 21
sub-Saharan Africa 205
subsidies and tariffs 196-8,
  abolition of 197-8
Suffolk, farming in 17

sugar beet and energy production 187
sugar cane 212
  organic 225
sugar farmer subsidies, EU 196-7
sugar, organic 231
Sumeria 165
sunflowers 206
sunhemp 231
'superbugs' 102
supermarkets 189
'superweeds' 116-7
Surveys of Pesticide Residues in Food 94
sustainability 32
sustainable agriculture (definition) 12, 87
sustainable energy 187-8
sustainable farming 194
sustainable systems 206, 207, 208
sustainable technology 24
Sweden 187, 188, 216
sweet potatoes, GM and non-GM 110
  Mexico 191
  weevils 228
sweetcorn 90, 201
swine fever 141
Swiss Federal Research Station for Agroecology and Agriculture 62
Swiss government support for farms 197
Switzerland 185, 209, 216, 219-20, 221, 230
  research into composting 137
Sykes, Friend 47, 55, 56-57, 59, 60
symbiosis, plant 68-70
synergy, whole-system 201
T25 GM maize 117-118, 123, 127
Tanzania 226, 230
Tao of Physics, The (Capra) 162
tariffs 196
Tauxe, Dr Robert 92
Taylor, Arnold 130
tea, organic 225, 226
terra mater 173
terra preta ('black earth'), Amazon, Brazil 232, 233
Texas 222
textiles 41
Thailand 152, 205 221

rice farming 200
Third World 10, 13, 14, 152, 184, 211, 227
  agriculture 87, 112
  agriculture, future of 205-9
  and GM crops 31, 108, 124
  and organic practices 110-1
  and pesticides 93, 100
  and population control 184
Third World farming 10, 208-9
  and EU subsidies 196
  and Western methods 206
thrush (song) 101
tolclofos-methyl 94
Tolhurst, Iain 200-1
topsoil 78
Townshend, Viscount 43
toxic chemicals 221
toxicological tests 126
'traceability' of food 190, 204
Trade Related Intellectual Property Rights Agreement (WTO) 129, 130
traditional farming 205
training, agricultural industry 193-4, 216, 221
tree planting 226
tribal peoples and Nature 159-61
Trichoderma 79, 228
Truth and Science (Steiner) 52
Tull, Jethro 36
Turner, Newman 10, 47, 57, 59-60, 70, 81, 82, 85
  composting 137
Turning Point, The (Capra) 191
Uganda 110, 226
United Kingdom 31, 42, 46, 80, 118, 141, 142, 143, 180, 185, 216, 230
  changes in waste disposal 143-4
  choice of pesticides available 23
  composting 136, 138, 140
  energy production 187
  food supplies 189-90
  forestry policy 203
  and GM plants 120, 122-3
  government and organic farming 237
  organic farming in and

public opinion 217-8
United Nations Development Program (UNDP) 35
United Nations World Food Programme 110
United Nations 205, 211
United States Compost Council (USCC) 144
United States 28, 31, 57, 80, 118, 117, 144, 148, 180, 209, 230, 231, 236
  agriculture 37
  agricultural initiatives 220-3
  and composting 138
  crop rotation in 43
  farming in 179, 191
  farm subsidies 196-7
  as food exporter 184
  and GM crops 31, 113, 120, 124
  and landfill 132
  maize growing in 210
  national organic standards 221
  patent laws and GM crops 129
urine 199-200
US cotton farmers 196
US Department of Agriculture (USDA) 221
  organic label 221, 222
US Environmental Agency 145
vaccination, livestock 81
valerian 80
vegetable boxes 200
vegetables 41, 48, 78, 90, 91, 94, 147, 152, 191 203
  organic 230
Verticillium 228
veterinary drug residues 203
Vietnam 79, 180
  rice farming in 200

Virgil 163, 164
Vitamin A 112, 113
Washington Department of Agriculture 223
wasp, predatory, including chalcid, as pest control 81
waste collection authorities 141
waste management societies 141
Waste Products of Agriculture: their Utilization as Humus, The (Howard) 52
Waste Strategy 2000, UK 143
water depletion 212
water hyacinth 11
water pollution 100-104
water power 187
waterglass 80
weeds and agriculture 77, control of 85-86
West Indies, agriculture in 50
West, industrialization of 155-6
Western agriculture, future for 189 ff
Western farming 14
Western scientific definitions 234
Western views of Nature 82, 167 ff, 179
Weston, Sir Richard 42
wheat 40, 83, 84,
  GM 130, 179, 212, 228
  comparisons, organic and conventional 195
whitefly 81
  control 228
WHO *see* World Health Organisation
wildlife, loss of habitat 91

wind turbines 186
windpower 187
windrow system 139
Windsor beans 40-41
wine 97
winter crops 41
Wolfe, Professor Martin 84
women and agriculture 111, 112, 205, 206, 227
wood as fuel 187
Woodland Grant Scheme 202-3
Woodlands Farm, Lincolnshire 218
Woods Hole Research Center, US 128
Wookey, Barry 85
Wordsworth, William 176-7
Working Party on Pesticide Residues, UK 95-96
World Bank 35, 205
World Health Organisation (WHO) 35, 102-103
World Resources Institute 99
World Trade Organisation (WTO) 125, 129, 205, 209
worm composting units 228
wormeries 136
worms *see* under type
wormwood 80
WTO *see* World Trade Organisation
Xenophon 210
Yorkshire 17
Yorkshire, farming in 17
Zambia 226, 227
zero-tillage 26-27, 85-86, 146, 147-8, 183, 198, 199, 200, 234

Also available:

SEEDS OF DECEPTION
Exposing Corporate and Government Lies about the Safety of Genetically Engineered Food

Jeffrey M. Smith

"This is a brilliant book . . . it positively fizzes with the human drama of the cabals and conspiracies behind the scenes which have littered the history of Big Biotech in its frantic efforts to get itself accepted. It is meticulously documented and powerfully written, somewhere between a documentary and a thriller."
——from the Foreword by Michael Meacher MP

256 pages   ISBN 1 903998 41 7   £9.95 paperback

For our complete list of books, visit www.greenbooks.co.uk